Cambridge Elements ☰

Elements in the Philosophy of Biology
edited by
Grant Ramsey
KU Leuven
Michael Ruse
Florida State University

THE CAUSAL STRUCTURE OF NATURAL SELECTION

Charles H. Pence
Université catholique de
Louvain

CAMBRIDGE
UNIVERSITY PRESS

CAMBRIDGE
UNIVERSITY PRESS

University Printing House, Cambridge CB2 8BS, United Kingdom

One Liberty Plaza, 20th Floor, New York, NY 10006, USA

477 Williamstown Road, Port Melbourne, VIC 3207, Australia

314–321, 3rd Floor, Plot 3, Splendor Forum, Jasola District Centre, New Delhi – 110025, India

103 Penang Road, #05–06/07, Visioncrest Commercial, Singapore 238467

Cambridge University Press is part of the University of Cambridge.

It furthers the University's mission by disseminating knowledge in the pursuit of education, learning, and research at the highest international levels of excellence.

www.cambridge.org
Information on this title: www.cambridge.org/9781009100328
DOI: 10.1017/9781108680691

First published 2021

A catalogue record for this publication is available from the British Library.

ISBN 978-1-009-10032-8 Hardback
ISBN 978-1-108-74169-9 Paperback

ISSN 2515-1126 (online)
ISSN 2515-1118 (print)

The Causal Structure of Natural Selection

Elements in the Philosophy of Biology

DOI: 10.1017/9781108680691
First published online: September 2021

Charles H. Pence
Université catholique de Louvain
Author for correspondence: Charles H. Pence, charles@charlespence.net

Abstract: Recent arguments concerning the nature of causation in evolutionary theory, now often known as the debate between the 'causalist' and 'statisticalist' positions, have involved answers to a variety of independent questions – definitions of key evolutionary concepts like natural selection, fitness, and genetic drift; causation in multilevel systems; or the nature of evolutionary explanations, among others. This Element offers a way to disentangle one set of these questions surrounding the causal structure of natural selection. Doing so allows us to clearly reconstruct the approach that some of these major competing interpretations of evolutionary theory have to this causal structure, highlighting particular features of philosophical interest within each. Further, those features concern problems not exclusive to the philosophy of biology. Connections between them and, in two case studies, contemporary metaphysics and philosophy of physics demonstrate the potential value of broader collaboration in the understanding of evolution.

Keywords: natural selection, causation, causalist–statisticalist debate, universality, metaphysics of science

ISBNs: 9781009100328 (HB), 9781108741699 (PB), 9781108680691 (OC)
ISSNs: 2515-1126 (online), 2515-1118 (print)

Contents

Introduction

Evolutionary theory has, seemingly since its inception, called for careful philosophical interpretation. Some reasons for this are straightforward enough: It brings into question our own place in the universe (Hull, 1986; Machery, 2008) and threatens the cogency of traditional theories of ethics (Street, 2006) and epistemology (Bradie, 1986). Other concerns fall more squarely within the province of the philosophy of science. As soon as Darwin had set the *Origin* to paper, it was quickly recognized that this was, at the very least, an unusual scientific theory. Its argument rests on a deft combination of analogy from sophisticated but still "outsider" knowledge like that of agricultural breeders or pigeon fanciers (Sterrett, 2002), cutting-edge geological theory supported by Darwin's own observations on the Beagle voyage (Hodge, 1983), and connections with a surprisingly rich philosophical background (Sloan, 2009). This did not look like ordinary nineteenth-century science, as Darwin was quickly told by figures as diverse as the physicist-astronomer John Herschel (Pence, 2018), the engineer Fleeming Jenkin (1867), and the geologist Adam Sedgwick (1860).

What kinds of worries might one raise when it comes to the nature of evolution qua scientific theory? First, evolutionary theory's relationship to the evidence that supports it – even with access to contemporary evolutionary modeling, theories of probability, and statistical inference – is quite convoluted (see, e.g., the debate over probability and statistics in Sober, 2008). Darwin relied upon both contemporary observations that could not necessarily be extrapolated cleanly to the case of long-term population change, and on the fossil record, which is often taken to ground very different kinds of explanatory enterprises (Currie, 2019). This basis has been supplemented by a panoply of further sources in the intervening century-and-a-half, ranging from the biochemical fundamentals of DNA to large-scale ecological observations that Darwin would have thought impossible. Such contributions increase confidence in our conclusions about changes in life on earth but certainly do nothing to make the theory–evidence relationship easier to comprehend.

The structure of Darwin's presentation of natural selection was also quite novel.[1] Building a conception of the generation of adaptations predicated on what we would now call a process of biased sampling across generations (Hodge, 1987), with only the rudiments of statistical theory having yet been developed, and very little application even of those to problems in the biological

---

[1] The *Origin* also includes Darwin's argument for common ancestry, which is less innovative in this sense, and was also much more rapidly accepted by biologists of his day.

world, is quite a feat indeed (Sheynin, 1980; Hacking, 1990). It left a number of open questions about how we should conceive of the organisms, populations, and species that Darwin's theory described. The modern answers to those questions would only start to take shape after nearly a century of debate and discussion concerning whether and how to integrate probability and statistics into our understanding of evolutionary biology (Pence, in press).

One of these open questions concerned the relationship between the process of heredity or inheritance and the kinds of explanations produced by natural selection (Gayon, 1998). How much knowledge of patterns of variation and its transmission from parents to offspring was really necessary in order to provide a suitable grounding for our conclusions about evolutionary change? Darwin had only a fairly weak theory to offer, and while Mendel's work appeared to provide an alluring possibility to complete the picture, figuring out exactly how Mendelian results – data, for one thing, only about the ratios of offspring produced by a single type of mating event, not natural populations – were supposed to provide explanations of Darwinian, populational, selective phenomena was by no means immediately clear.

A second such question – or perhaps better, a refinement of the first – addresses the importance of the underlying material basis of heredity for evolutionary theory. By the early years of the twentieth century it was increasingly accepted that, somehow or other, chromosomes were intimately involved in the transmission of characters. But what kind of account of them, if any, would be needed to finish the evolutionary story? Were Mendel-style phenomenological patterns in the production of offspring enough, or did we need to tie those down to their underlying physico-chemical details? Or, to reframe things from today's perspective, how much do the nature and transmission of DNA really matter to evolutionary biology? Unlike the first question, which seems somewhat anachronistic from our perspective, this cluster of issues is far from resolved, with debate still raging about the nature of mutations and their importance for evolutionary change (Merlin, 2016; Stoltzfus and Yampolsky, 2009; Stoltzfus, 2019).

Third, we have the related fact that variation seems to be a distributional phenomenon. Except in relatively isolated cases like the apparently clear-cut traits of Mendel's peas, variations are often normally distributed (Galton, 1889), and detecting natural selection in the wild was soon taken to involve (at least, by those who believed that there was such a thing as natural selection to be detected at all) examining the changes in these distributions over time (Weldon, 1893). This presented yet another disconnect between theorizing about natural selection and the "traditional" life sciences, whose morphological approaches seemed to have had no need for such methods thus far.

Fourth and finally – and intimately related to the last open interpretive question – we have the issue that most directly concerns my work here, one that, again, has remained with us to the present day. Natural selection is, in the end, a description of a *probable* outcome (namely, adaptation to environmental conditions). It offers us no guarantees: Fitter organisms are only, all else equal, more likely to leave more offspring than the less fit, and fitter traits are more likely to spread than the less fit. In that sense, it is unlike either a universal law of nature, taken to necessarily imply that, say, for any two bodies whatsoever in the universe, the force of gravitational attraction between them is and must be Gm_1m_2/r^2, or an inductive generalization inferred from observing a vast number of, say, white swans, ever on the lookout for a rogue disconfirming instance.

If this is so, then what exactly *is* natural selection? How should it be understood? As Hodge has described it, biologists seeking to define natural selection in textbooks on evolution readily agree upon a partial set of necessary conditions for natural selection to act: "variation, heritability of variation, and differential reproduction of heritable variation." All also recognize "that some further condition is necessary, so that once this is given there will be a set of necessary conditions that are jointly sufficient" (Hodge, 1987, p. 250). Put differently, we need variation, inheritance, and differential reproduction, but we must also delineate the right sort of differential reproduction – the selective kind – from others due to other kinds of influences on the evolutionary process – like that due to drift, migration, and so forth.

It is in making this final distinction that we reach a point of divergence among the textbooks. Some authors turn to statistical qualifications, namely that the differential reproduction at work in natural selection must be

> "consistent" or "systematic" or "nonrandom," all terms with no peculiarly biological content and drawn often from the terminology of statistics, while other authors insist that the differential reproduction must be due to differences in "fitness" or "adaptation," terms characteristic of the biologists', even formerly the theologians', lexicon, terms with an apparent teleological import. (Hodge, 1987, pp. 250–251)

That is, some authors appeal only to the statistical nature of the patterns of the variation on offer. Others argue that natural selection is picked out by its effects; it is the sort of process that produces increases in fitness or produces adaptations (it is this sort of explanation that Hodge rightly notes has a teleological flavor).

These two common moves are to be contrasted, he argues, with a third choice – namely, "an explicit definitional insistance on causation itself," for "differential reproduction in selection is distinguished from any in [non-selective cases like] drift by its causation" (Hodge, 1987, p. 251), in particular, by whether or not the variation at issue is causally relevant to the differential reproduction being explained. Here we focus neither on the statistical nature of differential reproduction nor on the outcomes that result, but on the (causal) nature of the *process* that connects variation to differential selection.

In short, Hodge sees three distinct ways in which we might interpret natural selection. We could attempt a purely statistical characterization, trying to specify in the mathematical properties of sampling from populations those that pick out natural selection, as opposed to other evolutionary sampling processes. Alternatively, we could look retroactively for the signal of natural selection in terms of its effects – though this runs a serious risk of reintroducing a sort of teleological explanation that Darwin has often been taken to have rendered unnecessary in evolutionary thought. Lastly, we could explore the nature of the causal connections underlying natural selection itself. If selection is constituted by a subset of the causal influences that impinge upon organisms in the wild, perhaps that subset shares a common feature that would let us pick out selective from nonselective change.

This Element is, essentially, an examination of the last several decades of efforts to describe the underlying causal structure of the theory of evolution by natural selection. As we will see, since the problem was first given a clear formulation in the mid-1980s, there has ensued constant and fairly heated debate concerning the appropriate way to understand the causal structure of selection. In Section 1, I will introduce what is now often referred to as the "causalist/statisticalist debate," with a particular focus on attempting to disentangle the question of the *causal structure* of natural selection from a whole host of other concerns that have since become embroiled in it. Section 2 introduces a novel way of presenting the debate, via diagramming these causal structures. The lens for introducing this structure will be the statisticalist view, but a more general version of that structure can readily allow us to represent the underlying causal commitments of several varieties of causalism, which is the task of Section 3. Finally, in Section 4, I'll use this new presentation of the players in this arena to try to make some advances, connecting to work on causal exclusion arguments in the metaphysics of science and philosophy of mind and on universality in the philosophy of physics.

While I certainly have my own position in this debate, having published on the causalist approach to selection, the aim of this Element is not to offer decisive arguments in favor of one position or the other. Rather, I hope here to be responding to a general sense of stagnation surrounding the discussion – arguments are drawn up, responses written, and yet the positions of few authors seem to engage in anything like systematic response to one another. The hypothesis underlying my approach is that this is in part, at least, due to our having pervasively talked past one another. One way, then, to try to repair the failure of our work to connect is to take up the task of clarifying exactly what is at stake. As I will argue in the next section, there are a plethora of interesting philosophical questions that have been lumped together under the banner of the causalist/statisticalist debate, and many responses to these, in turn, have been forced into the boxes of one "camp" or the other. My goal here is to take a first step toward separating such questions, to pick up one that I believe has been broadly under-studied in the literature, and to show that understanding it can provide profitable opportunities for connection with other areas in the philosophy of science.

One more caveat is due before I continue. I am explicitly narrowing my scope in what follows to a presentation of *natural selection*, with a few unavoidable detours into concepts of fitness, populations, and traits. Of course, to do this is to neglect philosophers' other favorite evolutionary process, genetic drift. It also sets aside the other half of the "big four" factors that would frame an introductory course in evolutionary biology, mutation and migration. It omits as well the wider array of influences on evolutionary trajectories whose importance we are increasingly coming to appreciate as calls for an "extended" evolutionary synthesis become more widespread (Pigliucci, 2007), including niche construction, development, and genotype-environment interaction (see, e.g., Uller and Laland, 2019). I am fully aware that some readers may be inclined to throw such an "old-fashioned" approach as the one I'm pursuing here across the room (if reading a paper copy). That said, I think the choice to restrict my attention to natural selection is defensible for two reasons. First, the kind of structural features of selection that I'll point out here will be equally relevant for any evolutionary process – that is, anything giving rise to population change as a result of aggregated, individual-level events.[2] Selection will thus, at worst, prove useful as an analogy for these other cases. Second, as we will see, the causal structure of selection has been the site of particularly heated debate in recent years, to which we should now turn.

1 The Contemporary Debate over Causation in Natural Selection

It is no simple matter to understand causation in evolving systems. Individual organisms have, since Aristotle, been paradigmatic examples of causal agents,

---

[2] When discussing various positions in the causalist/statisticalist debate in the first few sections, I will use "organism" and "individual" interchangeably, as has been common practice in this literature; I will return to this question and clarify the notion of "individual" important for my own analysis in Section 2.3.

but the onward march of the life sciences has both broken these organisms down into their component parts – organs, tissues, cells, proteins, and nucleic acids – and combined them into successively larger groupings – trait groups, populations, species, and ecosystems. Insofar as these are all descriptions of the same underlying organic phenomena, philosophers of biology have an interesting interpretive task in front of us: How do we harmonize these descriptions, and which ones ought we to prefer in which kinds of circumstances? Where is the real "causal action" in evolutionary theory? From Ernst Mayr's "population thinking" (1961, 1976), to the periodic heralding of the "return" of entities like groups (Sterelny, 1996) or organisms (Nicholson, 2014), to the focus of evolutionary theory, this issue has remained important to a variety of philosophers interested in the interpretation of natural selection.

1.1 The Received View

As classically crystallized by Elliott Sober's *The Nature of Selection* (1984), a "received view" of the fundamental causal structure of evolutionary theory was developed in the first few decades of contemporary philosophy of biology. On this picture, natural selection is a causal process driving evolutionary change. The strength and direction of this process are, in turn, described by Darwinian fitness. The proper way to understand this fitness property spawned its own literature, internal to the received view (Mills and Beatty, 1979; Beatty and Finsen, 1989; Brandon, 1990; Sober, 2001), too complex for me to engage with in great detail here. In short, the idea is that fitness describes not some simple, phenotypic or demographic property of an organism (like its speed, strength, or the number of offspring it has had to date), but rather its propensity to survive and reproduce – that is, how likely it is to have various numbers of offspring. Just as a fragile glass is likely to break into more pieces when dropped than a sturdy one, a fit organism is likely to have more offspring than an unfit one, holding all else equal. Fitness, on this approach, summarizes a vast number of the causally relevant properties of an organism that, when summed together in the right way, describe the probability of various future possible outcomes of natural selection.[3]

Natural selection is then taken to be a causal process operating on populations, the direction and strength of which are governed in part by fitness,

---

[3] Notably, this describes only the earliest forms of the received view; as we will see below, a number of more sophisticated versions of causalism are now on offer.

which is the propensity of an organism to survive and reproduce.[4] Darwin's key insight, on this reading, was to see that if an organism is more likely to survive and reproduce (or if a character trait makes the organisms that bear it more likely to survive and reproduce) then, in the long run and all else equal, that organism (or trait) will probably increase its representation in the population. Contra an objection most commonly ascribed to Karl Popper (1974), such a theory is not tautologous, because there is no guarantee of evolutionary success to be found here – the fittest are only *more likely* to survive.[5]

This view has a few convenient advantages. As Sober argued in his initial presentation (though the story is now recognized to be significantly more complicated than this; see Sober, 2013), it can cleanly separate the notions of "selection of" and "selection for." Some traits are directly promoted by natural selection, while others increase in prevalence only because they are correlated with traits that improve fitness. To borrow a classic example, hearts are evolutionarily favored because they pump blood, not because they make thumping sounds, even if the properties of pumping blood and making thumping sounds are in fact coextensive. There has thus been *selection for* pumping blood, while there has only been *selection of* making thumping sounds. The received view gives us the tools we need to distinguish these two cases. If a trait is associated with an increase in fitness in a counterfactually and causally robust kind of way, which we can discover by examining the propensity that describes fitness, then it is being selected for; if we can see that the trait is not in fact causally relevant to an organism's success, then it is merely benefiting from "selection of." We could imagine noiseless hearts that still contributed to survival and reproduction; we cannot imagine pumpless thumpers that do the same.

We can also use this approach to identify the differing causal impacts of the various "forces" or "factors" that contribute to evolutionary theory. Since selective evolutionary change is clearly defined, then other sorts of change – especially that due to genetic drift, which has been an important topic for philosophers of biology for other reasons (Hodge, 1987; Millstein, 2002, 2008) – can be examined comparatively. Net evolutionary change is thus considered to be the result of the cumulative impacts of each of these features, like selection, migration, mutation, drift, and so forth. This kind of

---

[4] This slippage between the organismic and population levels would go on to be one of the key drivers behind the development of more sophisticated causalist views.

[5] It is worth emphasizing that Popper himself eventually abandoned this objection (Elgin and Sober, 2017).

compositional approach (whatever ontology might in fact underlie it) matches extremely well with the way in which biologists talk about the process of evolution (Luque, 2016; Pence, 2017).

1.2 The Statisticalist View

The philosophical situation changed radically, however, in 2002, with the publication of two articles directly challenging this received view (Matthen and Ariew, 2002; Walsh et al., 2002). These articles inaugurated what has since become known as the *statisticalist* interpretation of evolutionary theory, by contrast with the received view, now known as the *causalist* position. The statisticalist approach is complex, and I will present it more thoroughly later. But it can be summarized roughly as follows: First, the statisticalists argue that natural selection and genetic drift (other factors in evolution are rarely considered) are merely convenient summaries of the genuinely causal events taking place in the lives of individual organisms. When organisms eat, fight, mate, and die, these causal events are what powers evolution. We, as theorists, then make the decision to abstract away from these fine-grained details and build evolutionary explanations using terms like selection, fitness, and drift.

This abstraction has several important effects, the statisticalists argue. First, the assignment of selection or drift to a given population change is relative to particular observers and their interests, and thus there is no single fact of the matter about which of these are occurring "out there" in the world (Walsh et al., 2002, p. 467) – which entails that natural selection and genetic drift (in the sense of the received view, at least) do not exist "out there" in the world, either. In this sense, natural selection is a bit like the Dow Jones Industrial Average: potentially extremely useful for us in understanding the stock market, but not an independent feature of the world to be discovered, and intimately connected to our subjective interests in continuing to measure it.

Second, statisticalists argue that these abstractions cannot be causal. At a sufficient level of generality (after, that is, enough abstraction), we are left only with a priori statistical identities. The effects of natural selection or genetic drift, that is, are only mathematical consequences of understanding populations as certain kinds of statistical objects. Multigenerational sampling of various sorts, as is required simply in writing down a formalism to represent an evolving system, entails certain kinds of changes over time, which directly correspond to the mathematical formalisms of selection and drift. Because these entailments are analytic, they are not causal.

Third, and as an immediate extension of this last point, models of natural selection are *substrate-neutral* – that is, they make no reference whatsoever to what it is that is being selected or what it is that is drifting.[6] The very arrangement of a population (that is, into individuals bearing traits with inheritance and differential success) will imply the existence of some kinds of change (namely, an analogue to selection, where better-reproducing types will be better represented in the future, and an analogue to drift, where the most probable result will be disrupted due to sampling error thanks to finite population size). If we arranged anything, from marbles to coins to elk, in those kinds of structures over time, we would see processes that looked very much like selection and drift in the "populations" that resulted.[7]

The extent to which this stands as a challenge to the received view should be relatively clear: Natural selection and genetic drift are not causal processes; fitness (at least, as we will see below, the kind of fitness actually relevant for evolutionary change) is not a causal property, much less a propensity; evolutionary explanations are subjective abstractions, not objective explanations of population phenomena; natural selection is indeed an a priori claim about certain kinds of statistical assemblages. Darwin should be viewed, they argue, as having described facts about the various causal events that individual organisms are involved in (instances of success or failure in the struggle for life) and their long-term consequences for individual lineages, which are *not* the target of the mathematical models of contemporary population genetics.

Such a challenge was rapidly picked up by the causalists, and what followed has been nearly twenty years (and counting) of ever-ramifying debate, encompassing more and more aspects of evolutionary theory. We have seen contention over the very definitions of natural selection and genetic drift (Millstein, 2002; Pfeifer, 2005; Brandon, 2006; Plutynski, 2007; McShea and Brandon, 2010; Ramsey, 2013b; Strevens, 2016), with a particular emphasis on a distinction between accounts of those that treat them as processes working in the world, and other accounts that treat them as identifying particular outcomes or results within populations (Brandon, 2005; Millstein, 2005; Millstein et al., 2009).

This has coincided with renewed debate over the nature of fitness. Authors have disputed whether or not fitness should be taken to be a property of

---

[6] This claim is not exclusively made by the statisticalists – for instance, it is very important to Sober (1984), a paradigmatically causalist work. It nonetheless forms an important part of the statisticalist argument, and so I will evaluate it as such in what follows.

[7] The presentation here – as with all my presentations of the statisticalist position in what follows – largely draws upon the recent "programmatic" piece by Walsh et al. (2017). I will, for reasons of space, be forced to pass over some important instances of differences of opinion under the broader statisticalist umbrella.

organisms, traits, or populations (Abrams, 2012a; Sober, 2013; Pence and Ramsey, 2015), whether the propensity interpretation itself is subject to a variety of proposed counterexamples and what changes to it would be necessary to resolve them (Ramsey, 2006; Abrams, 2009b; Otsuka et al., 2011; Pence and Ramsey, 2013; Ramsey, 2013c), and, perhaps most fundamentally, whether a causal property of fitness as proposed by the propensity interpretation can possibly fill the role that is demanded of it in the first place (Ariew and Lewontin, 2004; Bouchard and Rosenberg, 2004; Abrams, 2009a; Ariew and Ernst, 2009; Walsh, 2010; Ramsey, 2013a; Millstein, 2016; Triviño and Nuño de la Rosa, 2016).

Arguments over the nature of selection, in turn, have invoked the literature on causal processes, as well as Newtonian forces, in an attempt to clarify the locus of causation (Stephens, 2004; Millstein, 2013; Hitchcock and Velasco, 2014; Earnshaw, 2015; Luque, 2016; Pence, 2017). Note that this isn't quite the same as the classic "units of selection" debate (Okasha, 2006) – for one might believe that it is individual organisms that are being selected (that are the members of "Darwinian populations" sensu Godfrey-Smith, 2009), while still arguing that selection as a process acts upon populations. The question is more closely related to general considerations of supervenience and the relationship between organism-level, trait-level, and population-level causal facts (Reisman and Forber, 2005; Shapiro and Sober, 2007). As we consider how these questions are expressed in models of population genetics, we further have to understand the role of such mathematical models in the inferences drawn by population genetics, a topic that has been carefully explored by Jun Otsuka (2016, 2019).

Finally, one strand of the debate has looked at explanation as an approach to understanding the character of selection (putting a more epistemic gloss on a debate that has usually been framed in metaphysical terms), usually either pointing out the extent to which evolutionary explanations are sensitive to the context in which they are produced (Walsh, 2007), or in an effort to construct a new, noncausal sort of explanation that natural selection might involve (Matthen, 2009; Matthen and Ariew, 2009; Ariew et al., 2015).

There are, then, at least five different clusters of questions that all travel together under the banner of the "causalist/statisticalist debate." Placed roughly in order of decreasing proximity to biological practice, we have the following:

1. How should we define natural selection and genetic drift? Are they to be considered as processes acting upon populations, or population-level outcomes, or statistical identities? If they are processes, are they causal processes? Newtonian forces? Something else?

2. How should we define Darwinian fitness? Is it a property of individual organisms, or of traits? Is it a propensity (and hence capable of playing a causal role), or not? If it is capable of playing a causal role, what does that role actually look like?

3. Where is the "causal action" within an evolving population? At the level of individuals? Populations? Genes? (Notably, this could be independent of the questions in 1 and 2 if neither selection, nor drift, nor fitness are causal.)

4. What is the role of observer-dependence or abstraction in the generation of evolutionary explanations? Do evolutionary explanations take the form they do because of subjective choices made by researchers, and if so, what does that entail?

5. How should we understand the nature of supervenience and causal systems made up both of individuals and of populations of those individuals, as applied to the case of evolution? What role do mathematical models play in evolutionary inferences?

It is regrettable that, as the debate has progressed, positions on all five of these clusters of quite different questions have been lumped together into two overarching categories: "causalist" and "statisticalist." Not only does this make trouble for our capacity to evaluate those views – here lies, I believe, no small part of the dispute's ability to continue unabated for nearly two decades – but it also papers over interesting and relevant distinctions between positions otherwise given the same label.[8]

1.3 Disentangling and Crosscutting

In short, it seems somewhere between unwise and impossible to argue all of these questions at once. If this is so, however, we need to change our framework in an effort to move forward. I see two ways in which we might do that. On the one hand, we should remain aware that several of these questions can and should be dealt with in isolation. Some recent work on evolutionary explanations is a good example of this (e.g., Ariew et al., 2015), and discussions of fitness are increasingly disentangling the question of causation or ontology from the empirical question of measurement (e.g., Millstein, 2016). There remains ample room for the treatment of separate pieces of these five sets of questions, free from the burden of having to thereby support the rest of the broader "package" of the causalist or statisticalist positions. Clearly, the field

---

[8] I owe my appreciation of this point to extensive chats with Marshall Abrams, a friend and fellow causalist with whom I nonetheless disagree on a variety of interpretive questions. A book that he is currently preparing will, I believe, be another crucial resource for clarifying the current landscape of the debate.

would be well served by allowing for discussion of these issues *without* framing them in terms of this debate.

On the other hand, even if we accept that there is a vast diversity between these various questions, we might still look for possible relationships and connections between them. This would be a sort of via media between the current state of the art – that is, the "causalist" versus "statisticalist" dichotomy – and a fully atomized approach to each of these concerns taken separately. The idea would be to hunt for systematic ways in which a portion of the larger list above could be broken off and tackled in a consistent but isolated manner. This, by contrast with the above, is a fairly novel enterprise, in part because the diversity of issues in play here is so rarely recognized in the literature. With respect to questions about causal structure in particular, it will be the approach that I adopt in what follows. The main effort so far in this vein has been a recent book in this same series by Otsuka (2019). It's important, then, for me to briefly evaluate the nature of Otsuka's view before introducing my own.

For Otsuka, the task of interpreting evolutionary theory essentially involves understanding how it is that the equations of population genetics can successfully map onto the biological world. Since, as stressed by the statisticalists, one can frame these equations as statistical truths that hold a priori of any population with a given sort of structure (no matter its composition), if they are to be components of an empirical theory we owe ourselves a story about their applicability to empirical phenomena. For the statisticalists, as we have seen, this story is one of abstraction – as we make subjective choices to ignore certain kinds of biological details, we produce explanations that elide over those details, and these abstracted explanations, at their most general, invoke a priori mathematical identities.

This focus on the a priori character of the relevant statistical claims taken in isolation is, Otsuka claims, a mistake. When we investigate the full formalism required to derive the evolutionary trajectory of a population, we find that the relevant statistical identities only connect to one another in the right way if they instantiate a particular set of causal relationships in the world. These relationships, Otsuka argues, can best be described in the language of causal graph theory, which can both depict the formalism at issue and the causal connections that underlie it.[9] Rather than being merely a matter of analytic

---

[9] Though whether or not Otsuka's formalism has indeed captured the relevant evolutionary concepts is a matter of some debate; see McLoone (2018).

statistical entailment, then, "successful applications of evolutionary models are contingent on the causal structure of target populations" (Otsuka, 2019, p. 50). Only in the presence of the right kind of causal arrangement do the statistics gain any purchase on the biological world.

This presentation of Otsuka's work is too brief and has oversimplified his views (in particular, I am passing over Otsuka's rejection of the distinction between a priori and a posteriori components of scientific theories on broadly Quinean grounds), but it suffices to present the outlines of his approach. His work has been crucial for explicating and critiquing a key claim made by the statisticalists – namely, that the inferences underlying natural selection and genetic drift at their most general invoke a priori statistical identities and hence are analytic. But I will endeavor here to offer my own alternative framework, for two main reasons.

The first is that Otsuka's presentation, because it is focused on the application of the formalism of population genetics, operates almost exclusively at the population level. That means that the last cluster of questions I described above, on supervenience and the relationship between organisms and populations, fails to cleanly translate into his system. He notes as much himself and resists efforts like those of Shapiro and Sober (2007) to cash out the role of fitness in terms of supervenience. It is "the isomorphism of causal structure" at the level of populations, he argues, "and not just the supervenience of fitness values [i.e., the supervenience of fitnesses on the individual-level causal facts that ground them], that underlies extrapolation of evolutionary models, both diachronically and synchronically" (Otsuka, 2019, p. 52). I disagree, and I will offer a framework in which I think that disagreement can be profitably expressed.

Secondly, I'm searching here for a way in which to present this debate that would function independently of how it might be rendered in any particular mathematical formalism. Too much of the literature surrounding these questions is, I fear, phrased in terms solely of "saving the phenomena" of population genetics. This is an important enterprise, to be sure, but it is only a part of why understanding the causal structure of natural selection is so difficult. A new interpretive framework could, I hope, help us see more clearly when and where these various clusters of questions are connected, and when and where they are genuinely independent.

To sum up, then, Otsuka's treatment of these issues is vitally important and offers us a novel perspective on the nature of mathematical modeling in evolutionary theory. But there is still room, I think, for another way of parsing the landscape of the debate.

This alternative approach to the interconnections between these clusters of questions involves a more general view of the structure of evolving systems. To start to get leverage on such an enterprise, let's consider a fairly anodyne claim common to many debates in the philosophy of science: the separation of the ontology of a system from the behavior of the elements thus identified. Some philosophers have, of course, disputed that such a separation is in fact possible – particularly by arguing from a focus on scientific practice that we cannot understand the boundaries of our ontological categories without examining the uses to which we put the objects that are thus delimited, and hence ontology and system behavior are in constant dialogue as we build scientific theories (Love, 2018). Setting these worries aside for the moment, what do we find when we investigate the ontology of evolutionary biology?

One of the most striking features is that with which I began this section: its compositionality. As perhaps best exemplified by the perennial presence of talk of "levels of organization" in the philosophy of biology, and despite wide acknowledgment of serious problems with such a notion (Nicholson, 2012; Eronen, 2015), the idea that biological entities are arranged into causal hierarchies, from the cellular, to the organismic, to the populational, to the ecosystem, seems practically ineliminable from our biological and philosophical vocabulary. Efforts to understand that multilayered structure, then, could give us yet another way of cutting across these entangled clusters of questions, one that I hope will offer us both novel possibilities for smaller interventions – for tackling one problem at a time – and for engaging with philosophers outside the narrow remit of the philosophy of biology, perhaps finding some reinforcements in the struggle to understand causation in evolutionary theory.

2 Diagramming Evolving Systems

It is time, then, to develop some fresh tools for understanding the causal structure of evolving systems. In this section, I want to build a framework for thinking about exactly these sorts of questions, elaborated from a few initial insights in Walsh et al. (2017). Here, as a way to instantiate the general theory that I will offer, I will describe how we can use this new picture to understand statisticalism itself. In Section 3, then, I will turn to a presentation of two varieties of causalism in the same manner.

2.1 Two Kinds of Evolutionary Change

In discussing questions of causal structure, Walsh et al. (2017) draw a distinction that they take to be key for interpreting the statisticalist position:

The process of selection that Darwin postulated is in essence the change in *lineage structure* that occurs when there is variation in *vernacular fitness*. Accordingly, we shall call this phenomenon "Darwinian selection" (or "D-selection"). The other process, in which populations change in their *trait distribution* as a function of variation in their *trait fitnesses*, we'll call "Modern Synthesis selection" (or "MS-selection"). (Walsh et al., 2017, p. 4, original emphasis)

Let's unpack this. Walsh et al. describe a distinction between two different notions of fitness, operating at two different levels. First, we have a set of explanations at the individual level, where individual organisms change their *lineage structures* over time – that is, the character of the populations of offspring to which they will give rise (and, by extension, their relationships to the lineages of other organisms in the population). "As some organisms produce more offspring than others," they write, "their lineages increase their representation in the population" (Walsh et al., 2017, p. 3). While there is no further description of what exactly this change might look like (because, as they will soon go on to argue, it is not implicated in MS-selection, which is the primary target of their view, and hence not relevant), we could profitably read this emphasis on lineages in light of the model of the propensity interpretation of fitness proposed by Pence and Ramsey (2013, about which more in Section 3.2). On this picture, individual organisms are to be envisioned as giving rise to a branching tree of multigenerational lineages, some of which will be more successful – will result in an overall larger number of descendants – and some of which will be less successful. The future trajectories of individual organisms, then, are to be mapped in this sort of outcome space. Individual success is about doing better in the competition to produce larger lineages over time. Fitness at this level consists in what Walsh et al. call *vernacular fitness*, and corresponds essentially to the propensity interpretation as laid out above. It represents the sum of the kinds of facts about individuals in which Darwin was interested – being stronger, faster, and so on – that lead to individual success in survival and reproduction.

Second, we have a set of explanations at the population level, where populations change their trait distributions over time. These descriptions of evolutionary change, as typified by the models of population genetics, "provide us with a set of statistical parameters that explain, predict, and quantify changes in population structure" (Walsh et al., 2017, p. 2). On this view, population composition is expressed in terms of the relative distribution of traits: "populations undergo changes in their trait distribution; some trait types increase in their relative frequency with respect to others" (Walsh et al., 2017, p. 3). A population in this sense is fully characterized by some percentage of

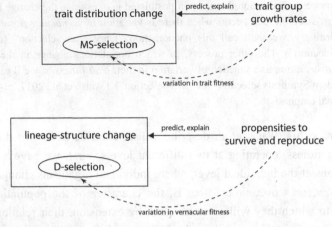

Figure 1 A diagrammatic representation of the two processes of natural
selection as described in Walsh et al. (2017).

its members possessing each of the traits that feature within it. Fitness at this
level consists in what they call *trait fitness*, here understood simply as the rate
of growth of a trait group within a population, in the sense familiar from the
formalism of population genetics.

Variation in each of these notions of fitness, then, is taken to give rise
to two corresponding concepts of selection. Vernacular fitness is used to
pick out a particular subset of lineage structure changes. Those changes in
lineage structure that can be traced back to differences in vernacular fitness
comprise *Darwinian selection* (while other kinds of lineage structure changes,
presumably, are nonselective). The same holds for *trait distribution changes*, of
which a subset, which arise as a result of differences in trait fitnesses, comprise
Modern Synthesis selection.[10]

The structure that Walsh et al. describe can be roughly represented by a
diagram like that in Figure 1.[11] Trait growth rates can be used to predict and
explain all sorts of trait distribution change. A subset of these, which are due to
variation of particular kinds in trait fitnesses, can be identified as MS-selection.
The same goes for D-selection, a subset of lineage changes that are the result
of variation in vernacular fitnesses.

Several things, I think, are striking about these two diagrams. First, which
will be important for further interpreting the statisticalist position later on, they

---

[10] Note that Walsh et al. say above that MS-selection arises "as a function of" trait fitness variation,
      indicating that they take this connection to be noncausal. I will return to this in the next section.
[11] I will carefully avoid referring to these diagrams or structures as "graphs" in the following, as
      they are not expressed in the formal apparatus of causal graph theory.

have exactly the same structure – even though, for the statisticalist, the top diagram represents a noncausal process and the bottom diagram represents a causal process. At the very least, that means that the correct way in which to interpret the arrows on these diagrams hasn't yet been established. They certainly should not all be interpreted as straightforwardly causal. As I will discuss further in Section 4.2, this question frames the statisticalist approach to the "substrate-neutrality" of evolutionary theory, an interpretive concern that lies at the core of the statisticalist view.

Second, and more importantly for my purposes here, these diagrams have a form that is familiar from across the philosophy of science, and particularly from literature in the philosophy of physics. On the left-hand side (the large rectangle), we have a state space of outcomes – either trait distributions or lineage structures – through which systems move over time.[12] On the right-hand side, we have a set of kinematic properties of a system (here, a population or an organism), changes in which produce motion through this state space. And in at least some specific cases, we can use certain kinds of dynamical principles (here, patterns of variation in fitness) that mark out certain subsets of trajectories of particular theoretical importance, the behavior of which can be systematically explained by the application of those dynamical principles – in this case, those corresponding to various notions of natural selection.

Just one ingredient is missing from Figure 1 to form a complete picture: the entities themselves. Individual organisms and populations are absent from the brief description of selection that we read above. Explicitly adding them back in is not just an exercise in idle completeness. Recall, in particular, that these two levels – individuals and populations – are connected by a composition relation. We won't be able to capture here the fact that populations are made up of organisms unless we give the entities being described a clear place in the analysis.

Doing so gives us Figure 2. We have entities that bear a certain set of kinematic properties. Those properties give rise to future property changes of the entities. Some of those future changes, in turn, may be explicable on the basis of dynamical principles defined over those same properties, which would then serve to pick out a subset of future trajectories of theoretical interest. Already, viewing the situation in this way helps us to see that there is something about this debate that transcends the disciplinary boundaries of evolutionary

---

[12] Strictly speaking, I will regularly describe this state space as being formed of trajectories instead of particular positions, because this is closer to the presentation of evolutionary change in the rest of the causalist/statisticalist debate. The two are, of course, harmlessly interconvertible.

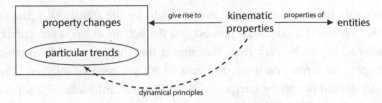

Figure 2 A generalized schema for interpreting the dynamics of a system.

biology. Further, it transcends even discussions of probabilistic or statistical theories – a diagram such as this could readily be instantiated in the context of a deterministic theory as well.[13] Before we unpack it further, however, a few disclaimers are in order, given that this diagram involves importing a number of terms that are not commonly found in the philosophy of biology.

Properties and Trends

First, it is taken as an assumption here that the sense in which entities bear properties is unproblematic, and hence we owe no further explanation of the "properties of" arrow in Figure 2.[14] Also, I take it that the "particular trends" in which we are interested at both the macro and micro levels are set independently, whether by our interests in building theories (i.e., they are the target of a specific class of explanations that we would like to offer) or by being the subset of property changes picked out by the dynamical principles, which we have identified independently. In our case, the particular trends at issue are various ways of instantiating natural selection – which we already, in some intuitive sense, believe to be the set of changes to individual or population trajectories brought about by variations in fitness and leading to adaptation. We thus have an independent handle on the phenomena we wish to explain; the devil is in the details of the causal story that gives rise to those changes.

Kinematics versus Dynamics

Second, on the distinction between *kinematics* and *dynamics*. Strictly speaking, *kinematics* refers to properties *concerning motion* – here, as is often the case in theoretical physics, the motion of an entity through a state space. To pick up our example, some kinds of changes in the properties of individuals will result in changes in lineage structures, and some kinds of changes in the properties

---

[13] This means that the question of causation in natural selection is not *merely* one of the interpretation of probability or the role of statistics.

[14] Of course, contemporary metaphysics has by no means settled the way in which this is supposed to work (Loux, 2006, ch. 1–3), but those complexities are thankfully unimportant for us here.

of populations will result in changes in trait distributions. The properties that lead to those changes are *kinematic properties.*

The *dynamical principles*, on the other hand, offer systematic reasons that help us understand movement through that state space. Models of MS-selection, in Walsh et al.'s sense, are taken to show us that certain kinds of variation in trait fitness (that is, certain patterns of change in the kinematic properties) pick out a set of future trajectories through the state space – that is, cases in which populations are undergoing MS-selection. The same is true at the individual level, where certain patterns of variation in vernacular fitness pick out sets of D-selection trajectories. In both cases, it is the dynamical principles (here, variation in fitnesses) that allow us to pick out the relevant set of outcomes. (The kinematic properties and dynamic principles proposed by the statisticalists will be described in much more detail in the next subsection.)

Causation

I have very carefully avoided any causal vocabulary thus far, and explicitly do not want to require a causal interpretation of either of the two left-hand arrows in Figure 2. Kinematic properties are simply those that *give rise to* motion through the state space, in some way or another. Such explanations could be causal, even mechanistic, or they could be noncausal, taking the form, for instance, of Ariew et al.'s *autonomous-statistical explanations* (2015). To use these different kinds of explanation would merely be to ground the connection between the kinematic properties and future changes in a different kind of way.

Give Rise to

Astute readers will notice that there is a difference between Figures 1 and 2 concerning the two "give rise to" arrows – that is, kinematic properties of individuals or populations are taken to give rise to future motion through their state spaces. Here, I have taken some liberties in generalizing the diagram originally offered by the statisticalists. For Walsh et al. (2017), changes in populations and individuals instead *predict and explain* trajectories through their respective state spaces.

My use of "give rise to," however, is carefully chosen. In general, the relationship between prediction and explanation is complex enough (Douglas, 2009). But further, and more importantly, whether or not changes in kinematic properties that give rise to motion through the state space will predict or explain that motion is not clear. If changes in kinematic properties are directly causally responsible for the future movement through the state space, then this would clearly provide an explanation for how those changes come about. Other kinds of connection less obviously yield explanations, however. If supervenience is involved (as I will argue below is the case for some versions of causalism), we may have causal-exclusion worries to surmount. If the connection is noncausal (as is the case in statisticalism), we might demonstrate that we have a robust noncausal explanation, relying, for instance, on structural features of the kinematic properties (in the biological case, most commonly statistical or probabilistic identities) to give rise to future property changes.[15] Or instead of such a complex, noncausal explanation, the connection might be a matter of simple analytic entailment or mathematical identity, if particular configurations of kinematic properties imply, a priori, that certain future changes will eventually occur. In short, the presence and nature of an explanation here is not a given.

Dynamical Principles

Next, what are the dynamical principles, and what licenses the way in which they connect certain kinds of changes in kinematic properties with certain kinds of resulting trajectories through the state space? Here, too, I want to be rather deflationary. Of course, if one instantiates this kind of diagram for physical systems, it may well be the case that they are genuine laws of nature, whatever it is that we might mean by that, and thus that the resulting trajectories are *guaranteed* to appear given the right kind of changes in kinematic properties.

But granting that the prospects for "laws of nature" in biology are famously poor (Beatty, 1993; though see critique in Lange, 1995; Haufe, 2013), we need no such natural necessity to ground the explanation of particular trajectories by dynamics in the biological case. While the standards for what should count as a "good" explanation are, as I have already noted, themselves part of what is up for debate between the causalists and the statisticalists, we can rest content for the moment with explanatory pluralism about, to repeat my earlier definition of dynamics, the reasons for which some particular (in our case, selective) trajectories come about.

A helpful way to understand these last two features is by comparison with Sober's distinction between *source laws* and *consequence laws*. On Sober's view, source laws tell us what kinds of circumstances will give rise to various particular types of evolutionary forces (Sober, 1984, p. 50) – for instance, a population exhibiting some given features will be subject to a selective

---

[15] While I lack the space to enter into the details here, I have in mind recent (and extremely exciting) literature on noncausal, mathematical explanation, which points out the significant and complex role of causally impotent mathematical structure in generating scientific explanation and understanding (see, e.g., Baker, 2005; Huneman, 2010; Lange, 2013, 2016; Andersen, 2018).

pressure, or a population exhibiting another set of features will experience genetic drift. Consequence laws, by contrast, describe how it is that resulting forces give rise to overall evolutionary change. The upper "give rise to" arrow, then, is the domain of consequence laws – they describe the full state space of possible future outcomes in terms of the set of all present kinematic properties, without necessarily offering detailed explanations of how those trajectories arise. The dynamical principles, in my sense, correspond roughly to Sober's source laws. These actually provide reasons for which a particular set of outcomes would take place as a result of particular arrangements of the kinematic properties.

2.2 Kinematics and Dynamics in Statisticalism

The primary kinematic property that Walsh et al. describe at the individual level, then, just is the propensity interpretation of fitness as we considered it above. Seen as derived from an underlying basis of individual-level causal events – "the causes of survival, deaths, immigration, emigration, reproduction, and inheritance of characteristics" (Walsh et al., 2017, p. 8), among others – it is "the propensity of *individuals* to survive and reproduce" that allows us to "explain or predict change in *lineage* structure" (Walsh et al., 2017, p. 3, original emphasis). Because changes in survival, migration, reproduction, and inheritance cause individuals to move through lineage-structure space, we can use changes in those properties to predict and explain future lineage structure changes. They are hence kinematic properties in my sense – they give rise to motion through outcome space.

In certain cases, they argue, we can specify some such changes in individual lineage structure as "selective," which they call D-selection. "The process of selection that Darwin postulated," they write, "is in essence the change in *lineage structure* that occurs when there is variation in *vernacular* [i.e., propensity-interpretation] *fitness*" (Walsh et al., 2017, p. 4, original emphasis). They do not speak, however, about just what kinds of dynamical principles would underlie this change – because they believe that there are no such general principles. There is, that is, no general way to move from changes in kinematic properties to a set of trajectories through lineage-structure space that should be called "D-selection." We can, of course, offer a D-selection–based explanation of some particular population change (like Kettlewell's darkening moths), they argue, "but this D-selection explanation is highly specific, and not generalizable. It does not tell us what would have happened if the specific history of births and deaths [of those moths] in Birmingham had been slightly different" (Walsh et al., 2017, p. 8). Bhogal (2020) helpfully frames this question in terms of what he calls *robustness* – the explanation that starts from population characteristics holds in more possible worlds (i.e., is compatible with a larger number of micro-level population constitutions) than the explanation that cites particular individual-level details.[16]

Such population-characterizing frequency distributions, Walsh et al. emphasize, are theoretical constructions, insofar as populations do not come naturally divided into their trait groups, and there are as many such classifications as there are traits of interest. In their words, in order to build these representations of populations and understand their change,

> we need to be able to divide the population into abstract trait classes (or "trait types"). We do this, for each heritable trait, by collecting together all those individuals that share that trait. The population of trait classes cross-classifies the population of individuals – each trait class has many individuals as members, and each individual is a member of many trait classes. (Walsh et al., 2017, p. 3)

These representations in terms of trait classes, in turn, will govern the correct set of kinematic properties that we need to predict and explain a population's trajectory through this state space. They are derived from trait frequencies and constitute "the predicted growth rate of the trait type in the ensemble. Only with knowledge of these growth rates can we predict and explain the change in a population's trait distribution" (Walsh et al., 2017, p. 3). Locally specified (i.e., for a given population) growth rates for each of the traits at issue are thus the right kinematic properties – where here these properties now depend not only on the way that we carve populations into traits, but also on the structure of the particular population. Working out these growth rates just is the business of specifying particular models of population genetics, ranging from the more abstract and theoretical (Rice, 2004) to the more practical and empirical (e.g., Ayala et al., 1974).

We are now in a position to identify what Walsh et al. call MS-selection and the dynamical laws that describe how it takes place. Most simply, MS-selection is the process "in which populations change in their *trait distributions* as a function of variation in their *trait fitnesses*" (Walsh et al., 2017, p. 4, original emphasis). As it turns out, the exact way in which to specify MS-selection trajectories as a subset of all possible trait distribution changes

---

[16] One could still attempt to demonstrate, contra Walsh et al., that the individual-level explanation is indeed generalizable in the right kinds of ways; this would lead in the direction of individual-level causalism (see Section 3.2). Taking Bhogal's suggestion that this kind of graded explanatory goodness implies the presence of objective deterministic chances, on the other hand, could produce an account quite similar to that of Abrams (see Section 3.1).

is left unspecified. This is likely because they consider the construction of precise models, which in turn specify selection dynamics, to be the province of population genetics.

But we can get some general purchase on the role they have in mind for these models by turning to their earlier work. As Matthen and Ariew write, "when there are heritable differences in traits leading to differential reproduction rates, the probability of the fitter types increasing in frequency is greater than that of the less-fit types increasing" (2009, p. 211). This focus on the outcomes for populations is echoed by Walsh (2007, p. 292), who writes that, on the statisticalist view, "selection-the-effect is explained by the inequality of $\Pr(p)$ and $\Pr(q)$," that is, the differing probabilities of types' increasing in frequency. An explanation that invokes natural selection, they argue elsewhere, is an explanation of "why, given the distribution of [traits] in the population sampled, the expected outcome is to be expected" (Walsh et al., 2002, p. 455).

These are thus the outcomes that will result when variations in fitness lead fitter types in fact to spread, as we expect they will – that is, when they aren't thwarted by other factors. If they are thwarted, and the most probable outcome fails to materialize, we have a case of genetic drift. As Walsh writes elsewhere, "the expected outcome of this process [of population change with variation in trait fitness] is some determinate change in the population's trait structure. When the change in frequency of types varies from the expected outcome, we say that the population is undergoing drift" (Walsh, 2013, p. 302).[17]

Crucially, the laws connecting trait fitnesses to movement through the state space of population trait distributions are simply mathematical predictions derived from these models: They are claims that assign probabilities to future trait-distribution states on the basis of current growth rates in each trait. Such inferences require only the use of statistics and probability theory and do not invoke any claims about the biology in a particular case. One would be able to infer the probabilities that are assigned to such future trajectories any time that one instantiated a multigenerational, population-sampling setup with the kind of structure that we see here – Matthen and Ariew (2002), for instance, describe their analogues in certain kinds of investment portfolios. In the evolutionary case, some of these outcomes will be identified by our models as the expected, "selective" outcomes, while the occurrence of others will indicate genetic drift.

---

[17] I will in general pass over the debate concerning the nature of genetic drift. It is here perhaps most important that we engage in careful dialogue with biological practice, which is famously difficult to decode in matters concerning drift (Beatty, 1987, 1992; Millstein, 2008). I think my approach could usefully illuminate drift as well, insofar as drift is yet another subset of population trajectories picked out by a different set of dynamical principles, but I am forced to reserve this for future work.

But the real connection between the kinematic properties and the state space is, the statisticalists claim, entirely mathematical in nature.

This has two important consequences, one that I will introduce now and one to which we will return in the next subsection. First, they argue, because the probabilities of future paths through the state space are just the result of the current state of the population plus some straightforward formulas from statistics concerning change and growth rates, those probabilities, and hence, the evolutionary outcomes, are *analytically entailed* by the corresponding trait fitnesses. The sense of "give rise to" at work at the population level, they thus claim, is a matter of purely mathematical inference.

If this is so, then *MS-selection is noncausal* – that is, all models of population-level natural selection in the tradition of population genetics are noncausal. Because models of population change, on this view, just describe mathematical consequences of preexisting population structures, they are not in the business (by definition) of picking out any causes, whether those causes operate on individuals or populations. If a model of natural selection is tantamount to describing what future population-level outcomes are most likely, merely as a result of the population's statistical structure, then that model has not thereby cited anything causal in order to do its job. Natural selection is not a cause of population change.

Their second claim has to do with the relationship between the individual and population levels – our next topic.

2.3 Connecting Individuals and Populations

As I have by now probably overemphasized, one of the distinctive features of the interpretation of causation in evolutionary theory is that it takes place, at least potentially, at both the individual and the population level (among others). This means that one single diagram, connecting properties of only one type of entity to change in only one state space, will be insufficient to enable us to capture evolution as a whole. We must, therefore, duplicate the diagram in Figure 2 and instantiate it for both individual and population properties.

With a macro level and a micro level clearly separated, it is time for one further disclaimer. Throughout what follows, I will be focusing exclusively on two-level systems, and when I use this picture to describe evolution, I will fix the micro level as the population and the macro level as the individual. The choice to restrict to two levels is a matter of tractability – as we will see, the number of interpretive questions raised even by two levels of interrelated change is significant enough to pose real philosophical problems.

The decision to limit my focus to individuals and populations, however, deserves further discussion. On the one hand, it is a response to the extant

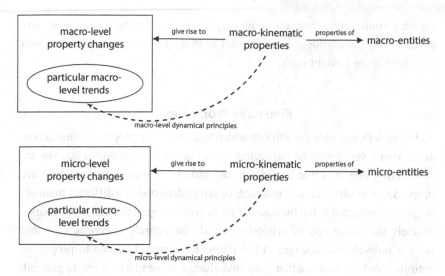

Figure 3 An uninterpreted diagram of a two-level causal system.

literature. The current debate, as we saw in the case of Walsh et al. (2017) above, is phrased as the interaction of individual- and population-level events, and so I will duplicate this framing for the rest of the work. On the other hand, I think that if we interpret the term "individual" sufficiently broadly – for instance, as "member of a Darwinian population" in the sense of Godfrey-Smith (2009) – then we might be able to generalize the kind of work that I perform here to other levels of selection, such as genes, groups, species, or ecosystems.[18] I will thus also endeavor in what follows to use "individual" in this more general sense (and thus neither in the sense of "organism" nor "biological individual," both familiar from elsewhere in the literature in philosophy of biology).

Defining the individual and population levels in this way, then, gives us our first arrow connecting the two levels: It is uncontroversial to note that the micro-entities *compose* the macro-entities – populations are made up of individuals.

Now, however, we have some real work in front of us. At the moment, the macro-level state space, kinematic properties, and dynamical laws are completely unconnected with their corresponding micro-level counterparts, and clearly this is an entirely unrealistic state of affairs. Equally evident, however, is the bewildering variety of ways in which we might understand connections between these two levels. I want to work through a number of

---

[18] I take it to be an open question whether such a generalization would actually provide us with an insightful way to approach the levels of selection problem; I lack the space here to consider that question in detail.

possible connections (both those actually advocated by the statisticalists and several other possibilities), in an attempt to imagine how various ways to knit together Figure 3 might work.

Kinematic Properties

To begin, let's examine the ways in which the kinematic properties themselves at the two levels might be connected. Here, in the evolutionary case, we are considering how it is that fitnesses at the individual level (whether those are propensities to survive and reproduce, or are understood in a different manner) might be connected to the fitnesses of traits within populations. It is, of course, unlikely that these sets of properties would be entirely unconnected – that is, it is unlikely that changes at the micro level would be due to properties entirely unrelated to those that governed change at the macro level. In general, a few different such relationships seem possible. First, we might see the macro-level properties as supervening upon the micro-level properties. In some cases, we might be able on this basis to offer a straightforward definition of the macro-properties in terms of the micro-properties – for example, if trait fitnesses are, as has been occasionally argued in the literature, nothing more than averages of the individual fitness values of the individuals that bear a trait. More-complex schemes of supervenience and multiple realizability might make such a clean definition impossible, while still retaining the same kind of general connection between levels.

We might also have more complex connections between these properties, which could in turn be described by more-complex structural or relational features of the macro- or micro-level systems than supervenience. For instance, one way of interpreting apparent phenomena of downward causation in the physical sciences, now often known as diachronic emergence (Guay and Sartenaer, 2016), describes them as cases where macro-level properties are constrained by the existence of temporally prior structural features at the micro level. Rather than involving any kind of "spooky" causal loops, we rather have micro-level facts at a given time t_1, which give rise to macro-level structures at time t_2, which in turn constrain the kinds of micro-level behaviors that are achievable at t_3, creating a kind of self-reinforcing emergent behavior. While such explanations are currently most often reserved for more-esoteric cases in the physical sciences, recent discussions of niche construction, for instance, have inspired the examination of reciprocal causation in evolutionary theory (Chiu, 2019), and it is possible that the kinds of explanatory resources developed in our understanding of emergence will be useful in helping us come to grips with these cases.

State Spaces

Next, let's consider the connections between the state spaces. While this is effectively never discussed in the literature in philosophy of biology, the possibility or impossibility of certain outcomes for individual lineages will certainly have effects on what kinds of population-level outcomes will be available – in other words, the structure of the micro-level state space and the structure of the macro-level state space will be linked. There are at least two ways in which we might think about this relationship. We could view the macro-level states as straightforwardly supervening upon the lower-level states. As with the case of the kinematic properties, the details here might be complex. For instance, how to translate facts about changes in individual lineage structure into facts about population change is by no means straightforward.[19] On the other hand, we might imagine examining connections moving in the other direction, as particular kinds of impossible population change (say, involving demographic properties or carrying capacities) constrain the available individual lineage changes.

Dynamical Principles

Lastly, we turn to the connections that might hold between the dynamical principles at the individual level and those at the population level. Such relationships might exist, though there has been very little consideration in the literature of how they might function. Some work on the propensity interpretation of fitness has aimed to forge such connections, allowing the derivation of certain kinds of population-level claims about natural selection on the basis of individual-level claims about propensities to survive and reproduce (Pence and Ramsey, 2013). Ramsey (2013b) has endeavored to construct a similar approach to genetic drift, where drift is quantified by a measure similar to the propensity interpretation of fitness.

In general, whatever one might think of this extant work, there are a number of ways in which we might imagine interconnections between dynamical principles. Macro-level laws might be formed, as we have already discussed, by abstracting away from detail present in the micro-level laws. This seems to be the approach favored by Matthen (2009), who argues that population-level evolutionary regularities result from our choices to abstract away from certain kinds of details in the lives and deaths of individuals.

---

[19] Adaptive dynamics is an interesting example of the sorts of complexity that will need to be engaged with in order to bridge such a gap (see Dieckmann, 1997; Kisdi and Geritz, 2010, for accessible introductions).

In the other direction, we might think that the behavior of micro-level dynamics is constrained by features of the macro-level dynamics. (At times, some work in adaptive dynamics seems to consider connections of this sort, imagining the behavior of a single organism exposed to a given set of population-level structures and selective pressures.)

As a third and final alternative, we might imagine that one set of dynamical principles was directly caused by the other (in either direction) – that is, it might be the case that macro-level dynamical features are caused directly by the micro-level dynamics, or, more controversially, we might have instances of downward causation, where micro-level dynamics are directly caused by macro-level processes.

The Statisticalist Alternative: No Connection at All

While these possible connections will be important for what follows, they are not relevant for our understanding of statisticalism – here enters the second major consequence of the analytic entailment between variation in trait fitnesses and future changes in population trait distribution. Since we would have such analytic entailments in any case in which we found enough of the right kind of structure to get a similar statistical sampling process (i.e., a chancy process of reproduction and inheritance) off the ground, a model of any such "population" would give us analogies to "selective" outcomes and "drift." This feature of evolutionary explanations Walsh et al. call the *substrate neutrality* of MS-selection. Put differently, population-level models simply do not care about the details of what happens at the individual level. "The information," as they write, "required to explain change in trait distribution is more or less irrelevant to the understanding of change in lineage structure and *vice versa*" (Walsh et al., 2017, p. 3).

The statisticalist position therefore claims to have an argument by which it can avoid all of the issues that result from the compositional link between the individual and population levels. If this composition is in fact irrelevant, because the behavior of the higher level is essentially independent from that of the lower level, then in fact the two-level diagram as laid out in Figure 3 has developed a sort of firewall across the center – the irrelevance of the lower level for the upper.

2.4 Statisticalism Summarized

With that, then, we can lay out the complete statisticalist view in Figure 4. As expressed here, we can see many of the view's most important features.

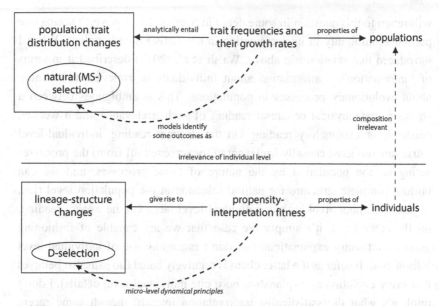

Figure 4 The statisticalist position, as found in Walsh et al. (2017), recast into the general framework of Figure 3. Italics indicate unspecified elements of the view.

The irrelevance of facts about individuals for natural selection separates the upper level from the lower, and converts the composition arrow into a fact about ontology that's rather meaningless from an evolutionary perspective. Further, the population-level connection between kinematics and state space is transformed here into analytic entailment. The privileging of MS-selection as one kind of dynamics within the broader state space is just a matter of certain outcomes being picked out by the relevant models, while the equivalent dynamical principles at the individual level lie outside the scope of the statisticalist view.

From the perspective of the causal structure of the statisticalist interpretation, we can see that many of the interesting properties of this view are wrapped up in what we might call the *isolation* of the top-left portion of the diagram from the rest. The connections between trait frequencies and future population growth are, as a result of the analytic entailment relations that hold between them and (thereby) the irrelevance of lower levels, insulated from any other causal facts taking place elsewhere. To put the same point a bit more provocatively, the single assertion that trait growth rates analytically entail population kinematics is carrying quite a bit of weight with respect to the rest of the causal structure of selection.

Finally, I should pause to clarify the nature of the irrelevance relationship that renders the individual level unimportant to the population level. While I will return to this question in some detail in Section 4.2, it is worth noting one particular ambiguity in its presentation in the statisticalist literature. When I introduced that relationship above, Walsh et al. (2017) described it in terms of "information" – information about individuals is irrelevant to learning about evolutionary processes in populations. This is ambiguous between a stronger, metaphysical or causal reading of such irrelevance and a weaker, epistemic or explanatory reading. On the stronger reading, individual-level causes are rendered causally irrelevant to (or screened off from) the processes acting on the population by the nature of those processes, and we can build a complete structure for natural selection at the population level (i.e., Figure 4) without invoking any individual-level facts. On the weaker reading, on the other hand, it's simply the case that we are capable of fashioning good evolutionary explanations that don't make any use of individual-level information. Insofar as the latter claim is relatively banal (no causalist believes that every evolutionary explanation *must* cite individual-level details), I don't think it's what the statisticalist interpretation intends, though some recent statisticalist work foregrounding the concept of subjective abstraction in the creation of explanations muddies the waters here somewhat (Matthen, 2009; Ariew et al., 2015). In much of what follows, I will presume the stronger interpretation.

2.5 Generalizing the Picture

But as I noted in Section 2.3, the claim that there are no connections whatsoever between the individual and population levels is a peculiarity of the statisticalist position. What kind of structure might we derive if we take the possibility of those connections seriously, something that seems necessary in order to capture a variety of causalist positions? Adding to Figure 3 all of the open points of philosophical interpretation that have been laid out in the preceding produces the significantly more complex Figure 5.

A reader would be forgiven for reacting with some degree of shock to this diagram, particularly after I have presented the proliferation of unsolved interpretive problems as a major obstacle for this debate in Section 1. How are we possibly supposed to make any advances in the current state of the field with the tangled diagram present in Figure 5 serving as a tool? Of course, we have already seen that this diagram arises relatively organically from enumerating a variety of alternatives to the statisticalist picture, and that statisticalism itself can be represented in the simplified version of the diagram depicted in Figure 4. In the next section, I will offer a reinterpretation of two different causalist positions in exactly these same terms and point toward several novel

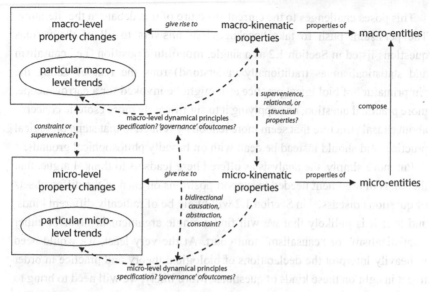

Figure 5 A two-level causal system with a number of potential loci of philosophical interpretation noted. Items in italics will, at least in some cases, require the resolution of difficult or controversial questions.

connections with related problems that this framework will allow us to more easily see. But before I do so, it's worth emphasizing that the theoretical situation is not as dire as it might seem.

One of the major concerns with the clusters of questions that I laid out in Section 1 was the fact that they appear, at first glance, to merit independent answers. Some kinds of questions that are raised there deserve to be settled in close dialogue with the conceptual basis of biological practice. For instance, it seems impossible to understand how to define the concepts of natural selection, genetic drift, and fitness without a clear understanding of how those notions are used by practicing biologists. Of course, biological concepts will be fuzzy, but in order for us to be confident that we have indeed offered a conceptual analysis of natural selection (say) as opposed to some other notion, we will have to show that philosophical work connects to practical concerns in the right kinds of ways.

At the same time, biological practice will not be responsive to the demands of *every* kind of philosophical question we might pose. Attempting to derive clear answers on the basis of practice about causal analysis, the relationship of the individual level to the population level, or even scientific explanation itself seem progressively less likely to succeed.[20]

---

[20] This is a somewhat controversial position; several authors have attempted to do precisely this (Haug, 2007; Millstein, 2008).

This poses challenges to the current structure of this debate in the literature. The persistent push to take "packages" of answers to all of the various questions listed in Section 1.2 as a single, monolithic position (i.e., causalism and statisticalism as traditionally understood) runs the risk of taking the "imprimatur" of biological practice as it might be invoked for a narrow-scope, more practical question, and applying it to significantly more esoteric concerns about causal structure that seem more tenuously related to that same biological practice, and should instead be dealt with on broadly philosophical grounds.

Put more simply, the analysis as offered here leads us to think, I argue, that the kinds of argument needed to support positions on each of the different sets of questions discussed in Section 1.2 will likely be of radically different kinds, and thus it is unlikely that we will find a single argument in favor of either "statisticalism" or "causalism" tout court. At the very least, we would need to heavily interpret the declarations of biological theory and practice in order to get insight on these kinds of questions. More likely, we will need to bring to bear independent theorizing about the more general questions, in concert with an understanding of biological theory. It is precisely that kind of independent, nonbiological theorizing, particularly when it concerns notions of causation and explanation, that I argue will be useful for us to move forward in the causalist/statisticalist debate, and one of my goals here is to offer a way in which to engage in such general theorizing productively.[21]

Do the various connections in Figure 5, then, amount to the same sort of potentially problematic mixing of issues as we saw in Section 1? I don't think so. Although there is a wide variety of concerns raised in Figure 5, I believe that we have made some headway toward separating the set of worries about causal structure and individual/population level relations from the rest of the broader causalist/statisticalist debate. It seems significantly more plausible that these are questions that need to be tackled at the same time, and that respond to a unified set of philosophical arguments and evidence. If relationships of constraint or supervenience hold between the individual and population levels, then these will need to be spelled out in concert to give us a unified picture of the causal structure of natural selection.

Views about these particular questions can, of course, be extracted from the broader work of various authors in the causalist and statisticalist camps. We've already seen how a statisticalist view can be thus reconstructed, and we'll see

---

[21] An important predecessor here is Haug (2007), who considers the impact of causal exclusion arguments on what he calls the "grounding question" – in terms of Figure 5, the relationship between the macro-level and micro-level dynamical principles. I will discuss Haug's approach further in Section 4.1.

two more examples of how to do the same reconstruction for a causalist picture in the next section. Evaluating these remains a tractable enterprise.

It is undeniably true that one problem to which I have alluded above is both genuine and potentially worrisome: These are the most general frontiers of the philosophy of biology, bordering on general philosophy of science and metaphysics. Insofar as this is the case, which is somewhat unusual for the causalist/statisticalist debate, we will be on uncertain ground. But the point that I hope to argue in the remainder of this Element is twofold. First, we have been making such arguments already, whether we have acknowledged it or not – positions on exactly these questions underlie both the causalist and statisticalist approaches as these are usually understood. Occasionally, the argument has been clearly and explicitly phrased in such terms. But more often than not, these positions must be carefully extracted from claims implicit in the ways in which causal connections are either defended or rejected.

Second, making those positions explicit and offering a framework by which we could begin to evaluate and understand them is certainly methodologically preferable to treating them indirectly, in a sometimes highly complex biological context. If one's primary goal is to defend particular interpretations of selection and drift, for example, or to take a certain approach to the mathematical formalism of population genetics, there is simply no guarantee that an argument for such a position will lead, inexorably, to a clear and straightforward position on these more general questions.

If we recognize, however, that these questions are indeed general, then they will form an excellent locus for us to find common cause with other areas in philosophy of science and metaphysics. I'll close this Element with examples of two ways that we might do that, in Section 4; but for the moment I want to end this section by emphasizing a few senses in which the generality of the questions in Figure 5 can be an advantage.

Related to a point made just above, one benefit of the recognition that these questions extend beyond the scope of the philosophy of biology is that we can clarify what kinds of evidence might be brought to bear on them. Analogies with other sciences and borrowings from general philosophy of science and metaphysics of science become open for us in ways that they have not been in the past.

While I lack the space to pursue them here, links also open up to the history of science. Worries about the construction and dynamics of multilevel theories have been with us since our first efforts to use statistical abstraction for scientific purposes, and I have engaged with some of this material elsewhere (Pence, 2021; Pence, in press). Perhaps more interestingly, and much less well understood, are connections with the historical development of statistical physics. Debates in the late-nineteenth century over the nature and characteristics of thermodynamic systems bear on exactly these same questions of multilevel dynamics and should be a fruitful source of further insight for the understanding of evolutionary theory.

To sum up, Figure 5 represents a novel way of looking at the interconnected causal structure of a two-level theory like evolution by natural selection. It's now time to use this structure to attempt to clarify two different examples of causalist positions.

3 New Perspectives on Causalism

Our next topic of discussion, then, is rather obvious. With statisticalism already laid out in Figure 4, how can we view the various causalist positions as specifications of the general structure laid out in Figure 5? Before I begin, I should preface this section with an apology; there are, as I have hinted at above, too many distinct positions that travel under the umbrella of "causalist" for me to be able to present them all here. I have aimed instead at selecting those exemplars that make claims about their causal structure clearly enough that I feel confident that the translations of them that I will provide don't do undue violence to their authors' original works.

3.1 Population-Level Causalism

Even more so than in the case of statisticalism, the causalist position turns out to be extremely diverse when we examine the ways in which it has been defended across the literature. I'll begin by reconstructing a view from two papers of Marshall Abrams (2012a, 2015). (I should note that Abrams denies [pers. comm.] that this position was in fact ever his at any particular point in time; we might thus call it a reconstructed "Abramsian" picture. His forthcoming book will present more thoroughly his current view.) This work is particularly helpful for my purposes, in that it specifies an interpretation of the underlying causal structure clearly enough that we can straightforwardly render it in the framework offered by Figure 5.

Abrams's broader goal is to use a theory of *causal probability* in order to attack the question of causation in natural selection.[22] Among all the sorts of probability that one might find in scientific practice (including at least subjective credences and, possibly, objective measures of indeterminism),

---

[22] I should note as well that Abrams has argued – in a manner that echoes my claim here that neglect of the general causal-structure question that I am exploring has hampered this debate – that failing to adequately consider interpretations of probability has been an equally systemic and problematic shortcoming of the literature.

causal probabilities are marked out in virtue of having a special feature. In Abrams's words, "altering the values of these probabilities by altering features of the chance setup that realizes them" – that is, following Hacking's coinage, the repeatable structure in the world that gives rise to multiple chance trials, like the experimental setup for flipping a coin – "would, in a certain sense, manipulate frequencies of events in the world" (Abrams, 2015, p. 529). If we then follow a manipulationist understanding of causation (e.g., Woodward, 2003), these probabilities can thus be said to play a causal role in the future occurrence of their outcomes.[23]

If evolutionary theory can be interpreted such that the outcomes of natural selection are described by causal probabilities, then, a fortiori, natural selection is a cause operating on populations, as it is a manipulable feature giving rise to population outcomes. Let's see how Abrams's support for this claim goes. I'll start by introducing his results concerning selection in particular, and then sketch (though not fully demonstrate here) the technical apparatus and arguments that underlie them.

To see the role that the conception of causal probability plays in Abrams's argument, let's start with (in the terminology of Figure 5) the connection between the micro-level kinematic properties and the macro-level kinematic properties.

> If there are ("lower-level") probabilities of factors that affect outcomes for individual organisms, these probabilities would mathematically imply probabilities of outcomes for the entire population. [... Further,] if the lower-level probabilities are causal probabilities then the population-level probabilities should be causal probabilities as well. (Abrams, 2015, p. 537)

Abrams thus sees an extremely tight connection between these two levels of kinematic properties. Causal probabilities at the individual level not only fix the values of the probabilities at the population level, they also fix the *nature* of those probabilities, in that they guarantee that if the former are causal in Abrams's sense, the latter will be, too.

This claim rests on two different formal results. First, probabilities at the higher level can, as a rule, be calculated from probabilities at the lower level. This is particularly easy if all of the relevant probabilities should be interpreted in the same way (for simple probability calculus is then deployed over the same kinds of properties) – but, Abrams argues, the point holds equally well in cases

---

[23] Such work forms an important part of a larger trend investigating the possibility of causally potent, macro-level probabilities that could function even in deterministic worlds; see, for instance, Lyon (2011); Abrams (2012b); Ismael (2009, 2011); Strevens (2011, 2013).

where we have mixed interpretations of probability, as causal probabilities abstract away from some of the features that might render us unable to perform such calculations (Abrams, 2015, pp. 534–537).

Second, because the properties of the lower level are straightforwardly also properties of the higher level, manipulating the lower-level properties in the higher-level system (i.e., manipulating individuals as part of a population) will cause changes in population-level outcomes. In other words, "causal probabilities in component setups imply causal probabilities in the chance setup composed of them" (Abrams, 2015, p. 535). If we can demonstrate that all of the individual-level probability distributions are causal probabilities, then we automatically know that the population-level probability distribution is causal as well.

This means, to sum up, that "the probabilities in terms of which natural selection could be defined are causal probabilities, if the low-level probabilities are, and in that sense, natural selection is a cause of evolution in a difference-making sense" (Abrams, 2015, p. 537). We can draw a clear arrow between the micro- and macro-level kinematic properties, here understood as causal probability distributions.[24]

Next, how might we see these causal probabilities as giving rise to the state spaces of individual organisms and of populations? About individual organisms, Abrams is not particularly concerned. He has argued elsewhere that fitness is not a property of individuals, and hence that the traditional understanding of the propensity interpretation is misguided (Abrams, 2012a) – there is thus no reason for him to consider the state space of individual property changes, as there is no fitness or selection to be found there. For populations, on the other hand, he argues for a supervenience view. "Natural selection," he writes, "supervenes on fitnesses, which in turn supervene on objective probabilities. These, in turn, supervene on whatever properties in a population of organisms constitute the relevant probabilities via an interpretation of probability" (Abrams, 2015, p. 523). The population-level causal probabilities we have just picked out, then, are the supervenience base of population changes like natural selection (and in turn are themselves to be understood by how they supervene on other properties of the population, should we wish to continue the analysis).

We're therefore left with one more arrow to fill in: the connection between the causal probabilities and the outcomes of natural selection, or, in my

---

[24] I leave aside here the further argument that the individual-level probability distributions are in fact causal probabilities; it is tantamount to the widely shared claim among both causalists and statisticalists that individual-level events like mating or dying are causal.

sense, the dynamical principles. For this, Abrams introduces some novel terminology: "We can define the word *govern*," he writes, "in terms of the relationship between causal probabilities and outcomes: When there are causal probabilities, the probabilities *govern* the outcomes" (Abrams, 2015, p. 534). While I lack the space to consider it here, Abrams turns to a case study from biological practice to demonstrate that "successful assumptions and practices embedded in contemporary evolutionary biology support the view that evolutionary outcomes are governed by causal probabilities" (Abrams, 2015, p. 544). In this sense, the governance relation is a reasonably informal one: It is simply what is said to hold when the outcomes of a system are described by causal probabilities.

With the final arrow drawn, then, we can lay out the "Abramsian" causalist position in Figure 6. In stark contrast to the relationship of irrelevance separating the two levels of the statisticalist Figure 4, we have both an important role for the fact that individuals compose populations, and a connection of entailment between the kinematic properties, fixing both their values at the population level, and the fact that they, too, are causal probabilities.

In terms of a general evaluation of this structure, we can see that much heavy lifting is being done by Abrams's theory of causal probabilities – it gives us the support for the three primary connections that differentiate this view. Interestingly, it is perhaps more radical than the statisticalist position in

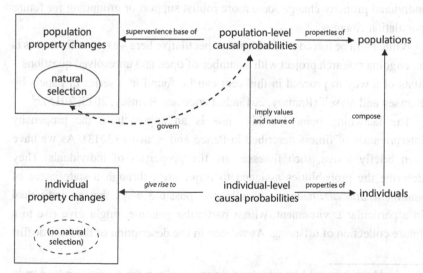

Figure 6 An "Abramsian" causalist position inspired by Abrams (2012a, 2015), recast into the general framework of Figure 5. Italics, as before, indicate elements left unspecified.

discounting the importance of the individual level. Rather than the presence
of a sort of "D-selection" process, analogous to natural selection but operative
at the level of individuals, we simply have no analogue of natural selection
operative whatsoever.

It is perhaps even a misnomer to pick out individual-level *kinematic*
properties in my sense from Abrams's work – insofar as there is no particular
dynamics being explained in the future outcomes of individuals on this picture,
one might argue that there is no real invocation of kinematics, either. That
would be a bit hasty, however. These are indeed properties of *change* in
individuals; in fact, they must be, given that they in turn ground properties
of population-level change.

We thus have a picture here that very cleanly picks out a single target for
selective explanations, at the population level, and describes the way in which
such selective processes are governed by a particular sort of population-level
probability distributions. These, in turn, are not sui generis at the population
level but arise from causal probabilities at the individual level, descriptions of
the chancy events in the lives and deaths of individuals.

3.2 Individual-Level Causalism

One might well wonder, however, whether a causalist view could be
constructed that would take more seriously the bottom-left corner of Figure 5 –
that is, a view that would attempt to find in the processes of long-term,
individual property change some more robust support or grounding for future
population changes.

While I will be forced to be a bit more speculative here – as we will see, this is
an ongoing research project with a number of open and unresolved questions –
hints of a way to proceed in this vein can be found in a series of papers by
Ramsey and myself (Ramsey, 2013a,b; Pence and Ramsey, 2013, 2015).[25]

The launching point for this view is an approach to the propensity
interpretation of fitness described in Pence and Ramsey (2013). As we have
seen briefly above, such fitnesses are the properties of individuals. They
describe the probabilities assigned to trajectories through a state space of
multigenerational lineages – that is, the possible ways that an individual
in a particular environment, with a particular genome, might give rise to a
future collection of offspring. As we saw in the description of the statisticalist

---

[25] I should note here that I have developed the presentation in this subsection independently,
and – particularly given the emphasis that Ramsey has placed on Dretske's distinction between
structuring and triggering causes in recent work (Ramsey, 2016) – I do not claim that he would
agree on the fine details.

approach to D-selection, we can think of fitness as parameterizing something like "individual success" in such a state space. If an individual is more likely to give rise to more successful lineages, it has a higher fitness; if it is more likely to die out or give rise to other less successful lineages, it has a lower fitness.[26]

That, then, fixes the micro-level state space and the micro-level kinematic properties, in the terms of Figure 5. Contrary to population-level causalism, however, this view goes on to argue that there is also an analogue to natural selection operating at the individual level. Because the propensity-interpretation fitnesses assign differential probabilities to the future changes in lineage structure, they can be taken to describe a process by which some lineages will be favored over time and others will be disfavored, leading to directional movement in this state space – in other words, a process of individual-level natural selection, analogous to the "D-selection" of the statisticalist picture. In Ramsey's words, "selection is the inter-organismic heterogeneity in the [state space of possible life histories] and is quantified via a function on this heterogeneity" (Ramsey, 2013b, p. 3914).

I will borrow a version (though expanded and more formal) of Abrams's coinage of "governance" for the relationship between the propensities described in propensity-interpretation fitness and individual-level natural selection. Governance here is not to be explained by anything more fundamental about propensities or the facts upon which they supervene – the fact that these individual-level probabilities over possible lineage structures will produce, in some cases, a future directional change (or a lack thereof) in this state space is nearly definitional, tantamount to saying that the propensity of a coin to land heads or tails governs the way in which it does so in sequences of flips, by summarizing the kinds of causal facts that are involved in coin-flipping.

One more micro-level arrow remains to be described on the individual-level causalist view: the connection between propensity-interpretation fitnesses and the state space as a whole (that is, beyond merely picking out the directional change of individual-level natural selection). Insofar as the propensity-interpretation fitnesses just are one way of tallying up certain facts about the larger collection of lineage-structure changes, it seems plausible to assert that those fitnesses supervene upon the set of lineage structures. This is paralleled by the way in which they are usually tackled numerically; we define a particular parameter across those life histories (often a complex function of

---

[26] I leave aside the question of how this property might be quantified; technical quirks quickly arise that fall outside my scope here. See Pence and Ramsey (2013) and, for important critique, Doulcier et al. (2020).

their size) and tally it up. The fitnesses thus calculated therefore arise from the broader set of facts about those lineage structures.

To move to the population level, we should first fix the relevant set of kinematic properties. Following standard practice, these are taken to be trait fitnesses. The connection between the propensity-interpretation fitnesses of individuals and the fitnesses of traits, on the other hand, has been hotly disputed. In the early days of the propensity interpretation, this was quickly set aside, as it was claimed that trait fitnesses could simply be taken to be the average value of the individual fitnesses for all the individuals bearing a given trait. After a pointed critique by Sober (2001) drawing on work by Gillespie (1974), however, philosophers have realized that such a definition leads to a number of unexpected consequences. One paper has argued that several of the arguments in favor of the fundamentality of trait fitness that Sober has raised (in the previously mentioned paper as well as Sober, 2013) do not in fact go through (Pence and Ramsey, 2015). This is only a negative conclusion, however, and there is as of yet no detailed theory of the emergence of trait fitnesses from individual fitnesses. As already mentioned, Abrams (2012a) has offered a number of arguments skeptical of individual fitness. The claims here are less than decisive on either side, and this is a location where future work is certainly to be desired.

How do individual-level causalists see the population-level process of natural selection as arising from trait fitness? While there is little clarity in the extant literature on this question, they have tended to make common cause with population-level causalists, arguing that, just as in the (strictly) population-level causalist view, variations in trait fitnesses underlie a causal process of population-level natural selection. That is, the same kind of governance relationship is posited to hold between trait fitnesses and future population change that was posited to hold at the individual level. (To say only this, however, is to leave this process underspecified – more about that in a moment.)

In much the same way that Abrams described population-level causal probabilities as arising from their individual level counterparts, for instance, Pence and Ramsey argue that a theory of selection, insofar as it needs to "cash out selection in terms of relevant physical differences between organisms," must be able to measure "how relevant a given physical difference is to an individual's reproductive success" (2013, p. 855) – that is, the process of picking out which causal influences underlie natural selection at the population level *just is* providing, in the end, an individual-level measure of fitness like that offered by the propensity interpretation. While, for perfectly good methodological reasons, we often do this at the population level via the intermediate step of computing trait fitnesses, there is no reason to think that we do not therefore wind up with the same relationship between individual and trait fitnesses as in Figure 6, with trait fitnesses being entailed by the values of individual fitnesses, in virtue of being grounded in the same difference-making properties of individuals.

Finally, we should consider the relationship between the trait fitnesses and the state space of population trait distribution changes as a whole. Once again, following Abrams's argument and the one offered above for the individual space of lineage-structures, it seems as though a supervenience relationship is the correct one here. Trait fitnesses supervene on particular elements of the broader set of facts about population trait distribution changes.

In short, the individual-level causalist position relies on a very strong analogy between the processes of selection happening at the individual and the population levels. In both cases, we have a state space of possible future outcomes. We have fitness measures that supervene on that state space that enable us to pick out particular kinds of "successful" outcomes within it. Those fitnesses, it is argued, do so by selecting the appropriate causal factors that give rise to the resulting success over time. As a result, we can consider them as governing a particular, selective kind of dynamical change within that state space. And, finally, all this is just as true of the individual level as it is of the population level.

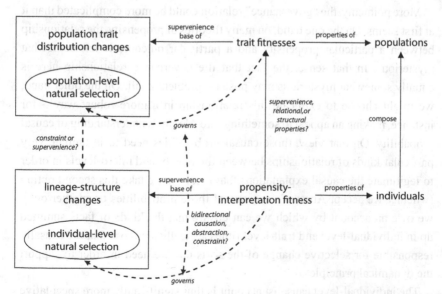

Figure 7 The individual-level causalist position, in the framework provided by Figure 5. Italics continue to indicate elements left unspecified.

Before moving forward, then, it's worth sketching the picture containing all the elements that we have specified up to now (Figure 7). The complete parallelism that I just noted is particularly striking when rendered visually – we have, in essence, a complete copy of Figure 5, only specified for the case of evolutionary biology. That said, there remains much work to do – there are several relationships that remain thus far unresolved on this view.

The main challenge for the individual-level causalist consists in tying the two levels together. It seems reasonable to suggest that the structures of individual lineage change and of population-level trait change both instantiate selective processes. It seems equally reasonable to expect that individual success and population success track one another – populations formed of better-performing individuals will themselves perform better, and better-performing populations are where we should expect to find better-performing individuals, all else equal. But we then owe an account that is capable of connecting together those two levels.

As already noted, individual-level causalists expect there to be some way to move from individual propensity-interpretation fitnesses to the fitnesses of traits within populations. Facts about the state space of future individual lineage change should be related to facts about the state space of future population-level trait distributions. And finally, the dynamical principles – the twin instances of "governance" – may well be related to one another in the sense that they are both giving rise to similar sorts of probabilistic dynamics. That leaves the individual-level causalist to fill in as many as three further arrows in Figure 7.

More pointedly, the "governance" relation could be more complicated than it at first seems. On the one hand, on many theories of propensity, the relationship between a particular propensity and a particular outcome is left somewhat mysterious. In that sense, the fact that the governance relationship here is equally somewhat mysterious may pose no problem at all. On the other hand, we might choose to explicate this relationship in a more robust way – for instance, making an appeal to something like Abrams's understanding of causal probability. On that view, those causal probabilities need to instantiate very particular kinds of relationships between the macro- and micro-levels in order to legitimate the causal explanations that result. If we take this second option (whether we accept Abrams's account of those probabilities or another one), we owe an account by which we can ensure that the kinds of facts summed up in individual-level and trait-level fitnesses really are those that are causally responsible for selective change of the sorts that we need in order to support the dynamical principles.

The individual-level causalist account is thus significantly more speculative than the other two alternatives that I have detailed here. It is not, however, often fully appreciated *as* a genuine alternative, and I think that one example of the utility of the approach I've developed here is the ease and clarity with which we can both describe and then indicate the open research questions arising from such a view. I will return to some of those open questions in Section 4.1.

3.3 Summing Up

While I have offered only three windows into the wide variety of efforts to depict the causal structure of natural selection in recent philosophy of biology, I think that these three are interesting enough, as well as different enough from one another, to provide a sort of proof-by-ostension of the utility of approaching the fundamentals of evolutionary biology using the structure I have described here.[27]

Much else concerning them will necessarily be left by the wayside. Returning to the list of entangled questions I introduced in Section 1.2, it is assuredly true that each of these views has been offered with associated definitions of selection and drift, each is implicated in discussions of the understanding of fitness, and each has important relationships to evolutionary explanation, all of which I have chosen to neglect here. Given that the three views are incompatible (and given the rather heated nature of the debate between them), all three are also armed with independent arguments against the cogency of the others. Dissecting these in their full detail would likely take more space than is available to me in this entire small Element.

More to the point, however, as I noted in the introduction, it is less my goal here to resolve this debate than to attempt to offer concrete proposals for how we might advance it – in particular, for how we might come to recognize its unappreciated reliance on precisely these more general (and generalizable) questions of causal structure. It is to these issues that I will now turn.

4 Moving the Debate Forward: Two Proposals

For the remainder of this Element, then, I want to consider how the framework that I have deployed might be useful not only as a way to clarify our thought about the causal structure of natural selection, but also to materially advance the debate over selection's causal status. Here I will investigate two such possibilities. First, if we more precisely evaluate the question of inter-level connections in the individual-level causalist picture, we will find resources in the metaphysics of science that can help us understand how the macro-level might be linked to the micro-level, focusing on the composition relationship.

---

[27] Perhaps most conspicuously missing from the causalist side, at least, are the views of Millstein (2006, 2013, 2016) and Hodge (1987). Placing this framework into closer dialogue with the work of Otsuka (2019) would also be a worthwhile effort.

Second, if we more carefully consider the substrate neutrality of evolutionary theory, we will find welcome common cause with philosophers of physics, who have explored similar problems in the application of physical theory. To anticipate a point I will make more thoroughly in the conclusion, recognizing that there is nothing uniquely biological about this portion of the debate – that is, the debate over causal structures – is perhaps our best hope for resolving it.

4.1 Connecting the Macro- and Micro-Levels

If we look at the kind of causal structure that is proposed by the individual-level causalists, we might be struck by a similarity with another traditional debate involving multiple-level causation: the question of the causal effectiveness of mental properties and the threat of "causal exclusion arguments," made popular in the philosophy of mind, most prominently by Kim (e.g., 1993). Evaluating this connection, made especially salient by the presentation in Figure 7, will again prove to be an interesting way to consider the question of the possible connections that could hold between the micro-level and the macro-level, and will allow us as well to engage with some recent literature on the relationship between causation and composition.

What, then, is Kim's causal exclusion argument? Imagine that we wanted to defend a nonreductive materialist view of the mind. We can think of nonreductive materialism as the conjunction of the following three claims. First, mental properties really are nothing more than the physical states of brains (and, hence, there are no souls or other spooky dualist stuff in the universe) – this is thus a variety of materialism.[28] Put a bit more precisely, and given that mental properties are multiply realizable (i.e., there is more than one brain state corresponding to the feeling of "being in pain," and the same for the rest of our mental states), we can say that fixing the state of a brain suffices to fix the set of mental properties currently being experienced, though not the converse. Second, and common to all sorts of materialism, brain states are responsible, at least in part (along with other internal and external influences), for causing future brain states. The particular brain state corresponding to being in pain might cause a reaction to move one's hand away from a flame, for example. And third, mental properties are causally effective as well – hence the claim that this materialism is nonreductive. For instance, it is correct to say that my being in pain caused me to be in a bad mood for the rest of the day (that is, my

---

[28] They are properties of brains as well as, to avoid taking a position on questions of extended or embodied cognition, any other parts of the body or environment that need to be fixed in order to fully determine mental properties. I'll abbreviate these to talk of the brain in the following for brevity.

mental properties caused my later brain states). When I say this, I'm not merely uttering some abbreviated version of a reductionist claim that only refers in fact to brain states – mental properties really are causal.

Kim-style exclusion arguments note that those three claims form a contradictory set if one adds to them what one might think is a fairly banal kind of claim about the way in which causation works: namely, a rejection of causal overdetermination. The idea is simple enough: There shouldn't be two different, independent, and complete causal pathways responsible for the production of the same effect. If we have, for example, at the brain-state level a complete causal story for the connection between my current being-in-pain brain states and my future being-in-a-bad-mood brain states, there's no *causal* work left for an account in terms of mental properties to do. On the one hand, this rejection of overdetermination should be welcome to a physicalist philosopher of mind – it prevents a sort of parallel, overdetermined dualism on which causally effective souls happen to follow exactly the same twists and turns as physical brain states. But on the other hand, Kim argues that it's fatal for nonreductive materialism. Brain states are the only things that are *really* causally effective; mental causation is nothing more than an epiphenomenon.[29]

It's fairly easy to see how this kind of critique might be targeted at the individual-level causalist. We have one, individual-level explanation for certain kinds of lineage structure changes that, when aggregated, result in some population change. We also have a population-level set of causes at the level of trait fitnesses that describe that same population change. If both of these are indeed complete causal explanations of natural selection, and the anti-causal-overdetermination principle at work in the mental case is also applicable here, then it looks like we have a contradictory state of affairs.

I am, of course, not the first to point out the analogy between evolution and mental properties. Two prior discussions are particularly relevant, the first offered by Shapiro and Sober (2007).[30] They liken the statisticalist position to a sort of radical epiphenomenalism about mental causation, of a type sometimes defended by Kim for the mental case, and they then go on to defend – using a causal structure not too dissimilar from my Figure 5 – the claim that we can in fact ground a manipulation-based notion of evolutionary causation applicable to selection, as long as we recognize which manipulations are indeed those

---

[29] As Shapiro and Sober (2007, pp. 242–249) note, Kim's position is more complex than this, as he seems to change his mind several times on the question of whether and how mental properties are causally effective. I present the most extreme version here, as it has the closest affinities with the evolutionary case.

[30] Reisman and Forber (2005, p. 1121) also raise the possibility of this sort of analysis, though they don't pursue it in much detail.

that would be required to test for the presence or absence of natural selection. A second author to have seriously evaluated this connection is Haug (2007). Haug's target is what he calls the *grounding relation,* which concerns (recast into my terminology) the relationship between the dynamical principles at the individual and population levels. For Haug, the answer to this question, in turn, will depend on the nature of the population-level kinematic properties (as Haug describes them, the "population-level properties involved in the process of selection"; 2007, p. 434). If a particular kind of causal realization relation holds between the individual-level kinematic properties and the population-level kinematic properties, then no causal exclusion problem will arise, because there are not two separate causal processes at work; there is simply one causal process at the population level, which is realized by a simultaneous process at the individual level.

I am sympathetic to both of these solutions – that is, the manipulation-based approach of Shapiro and Sober (2007) and the realization-based approach of Haug (2007). The former nicely connects with recent emphasis in the philosophy of biology literature on the manipulationist notion of causation popularized by Woodward (2003), and the latter is closely related to the kind of causal structure that I have defended above (in Section 3.2) for the individual-level causalist.[31] Here, however, I want to turn instead to some work that takes more seriously the causal import of the *composition* relationship between the individual and population levels, something that neither of the perspectives currently on offer takes to be an important part of the story.

Let's start with a fairly brief but extremely trenchant presentation of multilevel causal systems found in Boyd (2017). (To be clear, his treatment is inspired by the mental causation case, but intended to be fully general; the specification of it to the example of natural selection here is my own.) In short, he wants to argue for the following position concerning the relationship between micro-level causal details and macro-level phenomena:

> It is possible to confirm that some phenomenon is aggregated from small things ... and that the aggregation of their causal powers is *sufficient to produce* that phenomenon's characteristic causal effects without being able to identify the exact [micro-level] constituents in question. (Boyd, 2017, p. 285, original emphasis)

---

[31] Haug explicitly rejects framing this picture in terms of supervenience, arguing that a supervenience approach is too loose to rule out certain kinds of undesirable inter-level connections. He would thus disagree with that aspect of my presentation above. The nature of that disagreement, unfortunately, would take me too far afield here.

That is, we will sometimes be in the right position to make three simultaneous claims about a multilevel system like natural selection. First, natural selection is the result of the aggregation of multiple, individual-level phenomena. Second, the causal powers of the phenomena at the individual level, taken in the aggregate (and, hence, at the population level), are sufficient to produce natural selection, itself a macro-level cause. And third, we can learn that this is so even in spite of the fact that we might not know the exact individual-level details that gave rise to a particular selective change. In spite of the significant epistemic challenge of obtaining the individual-level details of an episode of selective change – gathering the relevant data is extremely difficult, and there are many sets of individual details that could give rise to the same selective change at the population level – we can still be justified in asserting that populations have macro-level causal powers in virtue of the micro-level causal powers of individuals.

How exactly are we supposed to do this? Kim's exclusion argument looms. A priori, we might hope that there is some kind of exclusion principle operative, because such a principle would support our claim that there is no sui generis action at the population level, action that can't be cashed out in terms of individual-level causes. In Boyd's words (for the case of mental properties), "the causal sufficiency of [the micro-level] to produce, when aggregated, all the effects we know about does not, by itself, entail that there are not other, nonphysical factors at work. [...] What is the evidence that there are no such nonphysical helpers?" (2017, p. 285). Of course, a full exclusion argument like Kim's would do the trick. But we would then be right back in the same trouble for an individual-level causalist position, with natural selection ruled out as an apparently contradictory case of causal overdeterminism. As Boyd describes it for mental causation:

> This would rule out nonphysical mental "helpers" all right, but (as Kim insists) it would rule out nonreductionist materialism as well because the causal sufficiency of microscopic physical causes to produce all natural effects (which nonreductionist materialists accept) would rule out the causal efficacy of macroscopic physical causes. (Boyd, 2017, pp. 285–286)

What we need, then, is a sort of "Goldilocks" replacement exclusion principle – one on which we can understand the connection between individual-level and population-level events without being forced either to discard one level in favor of the other or to permit causal factors other than those implicated in our individual- and population-level stories. As Boyd writes, "nonreductionist materialism says that macroscopic things are real and that big composite things can cause stuff *by being* composite. Still nonreductionist materialists *do* need a nonredundancy principle: composite physical things and their macroscopic powers are okay; explanatorily unnecessary dualistic posits are not okay" (Boyd, 2017, p. 286, original emphasis). What is this principle supposed to look like?

Here, Boyd claims, we need to carefully evaluate the characteristics of the particular explanations in play – and this is, he writes, always true when we need to consider a case where multiple explanations of the same phenomena are evaluated to determine whether or not they compete with one another. When we consider the value of "debunking" explanations in ethics, for example, we have to figure out whether some purported empirical facts undermine or in fact support a putative ethical explanation of the same phenomenon; this is not necessarily a simple prospect (Sturgeon, 1992). The same will hold for the case of causation. "There are cases," he writes, "where such non-competing explanations are mutually corroborative: sometimes the credibility of a macroscopic causal explanation is enhanced by an explanation of how the (more nearly) microconstituents of a macroscopic cause could, in aggregated concert, (help to) cause some macroscopic effect" (Boyd, 2017, p. 286). Whether or not the relationship will end in mutual support or undermining, then, is an empirical matter – just unpacking our notions of causation won't suffice.

Boyd thus argues he's given us a way to understand how these systems could be causal at multiple levels and nonetheless avoid exclusion, as well as how to evaluate when and whether a given macro-level cause is or is not excluded by the action of a given micro-level cause.

> So, when properly understood, the principle [of exclusion] does not rule out positing causal efficacy *both of* a macroscopic object and its macroscopic properties *and of* its [microscopic] components and their microscopic properties. Roughly, composite things and their composite causal powers do not compete with their components for causal efficacy *precisely because those components exercise the relevant causal powers by composing the composite object.* (Boyd, 2017, pp. 286–287, original emphasis)

The key move is to argue that the relation of composition is itself doing work here. It is precisely the fact that individuals are arranged population-wise that enables the population to be causally effective as a result of the causal powers of those individuals.

What would this look like in the case of natural selection? Here we would be arguing that the causal interactions of individuals – eating, mating, dying, and so forth – produce the causal profile of natural selection precisely when those individuals compose populations of certain sorts with certain kinds of structures and relationships. Those population explanations are supported and corroborated (rather than undermined) by the addition of individual-level detail demonstrating how they come to be realized. In some sense, what this does is provide a kind of "middle way" between saying that population-level natural selection is a mere analytic consequence of certain kinds of population structures, and the idea that there must be two independent selective processes at work, one at the individual level and one at the population level. On a Boydian picture of selection,[32] it is indeed a matter of definition that natural selection will occur given certain kinds of arrangements of individuals. Natural selection becomes something like a term for describing a type of those lower-level events, which they can realize in certain circumstances. But it is *not* thereby noncausal; rather, Boyd has demonstrated a way in which we might have a "definitional" connection that transmits causal efficacy from the lower level to the upper.

To return to the structure of individual-level causalism laid out in Figure 7, then, this way of understanding selection's causal efficacy amounts to a different kind of connection that could hold between the two relationships of "governance" at the individual and macroscopic levels.[33] For indeed, the fact that propensity-interpretation fitnesses govern selective outcomes at the individual level entails the fact that trait fitnesses will govern selective outcomes at the population level. To borrow an analogy from Jessica Wilson, to which I will return shortly, saying that a particular set of individual causal profiles gives rise to population-level selection is somewhat like saying that a given set of kinetic energies of particles and, by extension, a given mean molecular kinetic energy, give rise to the macroscopic phenomenon of temperature.[34] Indeed, this is essentially a definitional connection. But it is a definitional connection through which we can still see how the causal profile that we associate with temperature is produced in different cases *in virtue of* the causal profiles of each particle that give rise to the mean molecular kinetic

---

[32] I should emphasize that I do not wish to ascribe this interpretation of selection to Boyd himself; he approaches selection only briefly in Boyd (2017) and in a relatively different manner, concerned more with reductionism than causation.

[33] The emphasis on composition as playing an active role is similar to Haug's use of the realization relationship (2007), and is also related to Abrams's arguments about the connection between individual- and population-level causal probabilities (2015). But I think the idea that it is *composition itself* that is doing much of the work here is new to the context of this debate.

[34] It is noteworthy that this classic claim about the relationship between mean molecular kinetic energy and temperature is, strictly speaking, false. But it is nonetheless true that in statistical mechanics, temperature is "identified with a combination of more fundamental quantities," namely, the derivative of internal energy with respect to entropy (see, e.g., Skow, 2011, pp. 488–489). The calculation of entropy then leads to questions about ergodicity, which I will consider in the next section.

energy. At least in some cases, entailment via realization or identity like this does not have to imply lack of causal efficacy.

This still, of course, leaves open a fairly significant question: How do we *know* that we are in a situation like that of the relationship between mean molecular kinetic energy and temperature – that is, how can we be sure that the connection between a certain kind of micro-level structure and another kind of macro-level structure is such that it preserves causal efficacy in the way Boyd describes? Here, Boyd offers us no clear answer. To find one, we can turn to a discussion of compositional analyses as offered by Wilson (2010).[35]

Wilson's discussion centers around examples (similar to those discussed by Boyd) in which it seems that composition generates certain kinds of necessary connections between objects and aggregates. To take only the most relevant for our purposes here, she considers an identity or realization relationship – for instance, the claim that "anything having a certain mean molecular kinetic energy has a certain temperature" (Wilson, 2010, p. 197). Plausibly, this connection holds necessarily,[36] and the kind of individual-aggregate relationship involved in the temperature-energy case looks not entirely dissimilar from the analysis we considered of natural selection above, connecting a macro-level phenomenon to the micro-level details that give rise to it.

Wilson offers us four different criteria for the existence of a necessary connection that holds in virtue of a composition relation (Wilson, 2010, pp. 199–200). First, we know the actual, real-world causal profiles of having a given mean molecular kinetic energy and of having a given temperature – "that is, [we have knowledge] of what effects these properties, when instanced in certain circumstances, can enter into producing" (Wilson, 2010, p. 199). Second, the way that we tell these properties apart is by reference to those actual causal profiles. Third, the causal profile of having a given temperature is actually either identical with or contained within the causal profile of having a given mean molecular kinetic energy – "any effect that an instance of [temperature] can bring about in certain circumstances, in virtue of being temperature simpliciter, is an effect an instance of [mean molecular kinetic energy] can bring about when in those circumstances" (Wilson, 2010, p. 199). These claims, taken together, are just another way to say that "actually,

---

[35] As an aside, Wilson notes that the parts of her analysis that I will use here commit us to non-Humean accounts of causation. I will assume a non-Humean account in what follows and set aside those complications. In any case, she argues that the prospects for the Humean to explain the cases most relevant for my purposes are fairly poor.

[36] Wilson notes that one could resist the necessity of this and her other examples by pushing different metaphysical claims, but I won't follow her argument into those details.

anything that has a [mean molecular kinetic energy] has a temperature" (Wilson, 2010, p. 200). Fourth, and finally, these facts are modally stable – that they will continue to be true in other nearby possible worlds. "Such facts about a necessary overlap in modally stable causal profiles," Wilson writes, "provide a metaphysically straightforward and informative ground" for the truth of these identity or realization claims (Wilson, 2010, p. 200).

These four criteria could thus give us a way to respond to the objector who questions whether natural selection indeed instantiates the kind of compositional necessity that Boyd and Wilson have described. For such a connection to exist, it must be the case that (1) "being subject to a given selective pressure" and "having a certain, selective kind of individual-level lineage structures" are properties that have actual, known causal profiles; (2) we distinguish those properties by appeal to those causal profiles; (3) the causal profile of being subject to a given selective pressure is identical with or contained within that of having a certain individual-level causal structure; and (4) these facts are modally stable.

While this remains the subject of some controversy, biological practice seems to offer us sufficient grounds for the truth of claim (1). Working biologists are well aware of the effects of certain kinds of individual-level facts when taken in the aggregate, and how to characterize those facts as they bear on natural selection (see, e.g., Endler, 1986; relevant portions are discussed in Pence and Ramsey, 2015). We also know what it is for a population to undergo selection itself, and what kinds of effects natural selection will in turn have on the later composition of populations, on the environment, and so on. Claim (2) seems equally easy to satisfy on the basis of practice. Both selective pressures and individual-level lineage structure changes are features that we can identify (and separate from other evolutionary factors) on the basis of their empirical effects. In this sense, they are both quite similar to Wilson's invocation of temperature – we may well have work to do in standardizing definitions of natural selection and hence picking out particular selective pressures, as well as particular methodological or empirical questions to solve, but the property of being exposed to a given selective pressure will in the end be legitimated in much the same way as any other scientific property, as a stable feature of entities accessed by repeatable practices of measurement.

The last two claims, on the other hand, are more difficult to evaluate and deserve more space than I will be able to give them here. Is it the case that the causal profile of being subject to a given selective pressure is identical with or contained within that of having a certain kind of aggregated lineage structure? The statisticalist might argue that precisely what matters here is that details about the individual level are *thrown away* when we abstract our way to selection. We don't care about the "nature of the individual-level forces involved," as Walsh et al. write (2002, p. 463). But that process of abstraction is not necessarily a problem for this view. Temperature does not care about the nature of the particle-level collisions that give rise to mean molecular kinetic energy, either. We are identifying, on the contrary, structural features at the individual level – still the result of individual-level causal influences, but *after they are composed, and hence structured, in a particular way* – that give rise to the kind of causal powers that lead to natural selection.

The statisticalist might retort that what remains after this process of abstraction are only statistical features of populations, not causal ones – that is, that there simply is no causal profile for the property of being subject to a given selective pressure. But this would thus be an attack on claim (1) rather than an attack on claim (3). It seems difficult to reconcile the idea that in fact we do not know the causal profile of "being subject to a given selective pressure" with the accounts of practicing biologists. A reinterpretation of biological work could be possible – a few statisticalist papers purport to have presented such accounts for toy models involving coin-flips. But, in any event, we have shifted the target: This is no reason to doubt claim (3).

Perhaps the most promising line of attack comes from the statisticalists' interpretation of the models of natural selection. They argue that these are designed to "predict the magnitude and direction of evolutionary change" (Walsh et al., 2017, p. 4). Note the shift from *natural selection* to *evolutionary change* as a whole. Making such a shift entails that "extra information" beyond lineage structure is typically required, "including the effects of inheritance, details of reproductive schedules, mutation and migration rates, demographic factors such as the size and growth rate of the population as a whole, and the composition of variation" (Walsh et al., 2017, p. 4).

If this is true, though, then the analytic entailment at issue is not between trait fitnesses and *particular selective outcomes* – for now we have added to natural selection information about mutation and migration, population size, and so forth. Changes due to mutation and migration are, straightforwardly enough, not changes due to natural selection. This interpretation, that is, makes it seem as though the connection of analytic entailment that Walsh et al. want to discuss is rather the one between trait fitnesses and *the state space as a whole*. In the language of Figure 5, it is the arrow between the kinematic properties and the larger rectangle describing *all* of the possible population-level outcomes, including both selective and nonselective outcomes.

Once again, then, we do not in the end have a critique of claim (3). The claim that in order to produce the *entire* space of population outcomes, we need *more* than simply individual lineage-structure changes is surely true. But it is not an argument against claim (3), which argued rather that the causal profile of "being subject to a given selective pressure" is either identical with or contained within that of "having a certain individual-level causal structure."

Finally, and most difficult of all, is claim (4): the idea that this relationship between individual-level facts and population-level facts is modally stable. Will the fact of (3) continue to be true in other nearby possible worlds? While I confess that the exercise stretches my intuition, imagine what the failure of (3) to hold would look like. In such an example, we would have a population that exhibited something recognizably like the causal profile of being subject to a given selective pressure, which at the same time failed to exhibit the causal profile of having the equivalent sort of individual-level lineage structures that we would expect here in the actual world. But in such a world, what exactly is it that happens to the individuals that make up the population? How could we possibly get selection in favor of black moths and against white moths at the population level, without the lives and deaths of individual lineages being structured such that more black-moth lineages do well and more white-moth lineages do poorly?

At the risk of oversimplifying the point, populations are made up of individuals. We will not get frequency changes of populations without adding individuals of some types and removing individuals of other types. We will not get *selective* frequency changes of populations, in turn, unless those individuals are added and removed in ways that we recognize as selective (as opposed to by mutation, migration, random chance, and so forth). I find it hard to find a foothold for the rejection of claim (4).

Importantly, nothing in this subsection constitutes an objection to the statisticalist interpretation. Rather, I have argued that the ways in which causalists or statisticalists alike might contest the coherence of the model that I sketch here all seem to fail. That is, it seems that we have a genuine, live alternative in the form of a Boyd- and Wilson-inspired approach to natural selection as a fact arising through compositional necessity. Comparative evaluation of these projects – along with, at the very least, the population-level causalism of Abrams as described above – is a matter for another day.

4.2 On Substrate Neutrality

As I noted above, one of the most salient features of the statisticalist view, and one rendered all the more so by its presentation in Figure 4, is the fact that the top-left part of the causal diagram – the relationship between the trait frequencies and future population change – is "isolated" from the rest of the causal diagram. Neither detailed facts about the populations themselves nor about the individuals that make them up are taken to matter for natural selection at the population level.

I should note two important qualifications at the outset. First, whether or not those other facts are indeed irrelevant is a matter of debate. It is not clear, following the kinds of arguments developed by Millstein et al. (2009) or Abrams (2012a), whether this "isolation" is really a characteristic of natural selection or is simply a side effect of taking certain kinds of mathematical apparatus present in evolutionary theory (or, more accurately, within population genetics) and interpreting them outside their biological context.

Second, if one is familiar with the discourse around interlevel relationships in other sciences, the importance of substrate neutrality for the statisticalist position might seem strange. The independence of higher levels from details at the lower levels is often taken to be evidence in favor of the independent causal efficacy of those higher levels, quite the opposite of the statisticalist use of substrate neutrality to argue against population-level causation. I only have the space here to note this as an unusual feature of this debate worthy of further consideration. Those two caveats noted, this feature gives us further tools at our disposal to evaluate cases such as these, precisely because there is a long history of considering such relationships of relevance and scale in scientific theories.

The idea that a scientific theory might involve understanding the effects of micro-level phenomena at a higher level of organization, or in a limiting or asymptotic range of parameters, where this behavior gains a sort of independence from the behaviors of the system in other realms or at other levels, is by no means unique to the philosophy of biology. As Robert Batterman writes, once we have a basic understanding of the properties driving behaviors in a system, a straightforward next step in refining a theory

> is to see how much, and what types, of observed behavior can be generalized across distinct kinds of systems. Are there parameters that somehow characterize the dominant features of systems of widely different types? If there are, physicists often call the observed behavior and the dominant features "universal." (Batterman, 2000, p. 120)

There is thus an extensive literature on this concept of *universality* in physics, deriving especially from discussions of statistical mechanics. Equilibrium statistical mechanics manages to derive generalizable results about gases and fluids that are made up of impossibly large numbers (recall that Avogadro's number is of order 10^{23}) of radically different parts (i.e., with different internal constitutions, different interaction forces, etc.). And yet the kinds of macroscopic measures of those systems that statistical mechanics produces – temperatures, pressures, entropies, and so on – are sufficiently robust to characterize most of these systems extremely well.

If we could better understand when and how universality applies to physical theories, when we combine this with the kind of approach to causal structure developed here (that is, a more generalized approach to causal structure that is better able to be put in dialogue with developments in the philosophy of physics), we might be able to shed light on exactly the question of the "isolation" of population-level explanations that has been emphasized by the statisticalists. Criteria developed for examining physical universality could be used to clarify what certainly seems to be either universality or a closely allied concept proposed to be at work in the biological realm.

The fact that an analogy might be found between evolutionary theory and statistical physics is, of course, not news to anyone familiar with the history of evolutionary biology. R. A. Fisher, who spent a year just after his graduation from Cambridge working with the statistical physicist James Jeans, famously drew several such analogies in his 1930 work *The Genetical Theory of Natural Selection*, now considered one of the classic founding works of the modern synthesis. Much has been made of precisely this connection by statisticalist authors (including Walsh, 2003, 2007; Ariew et al., 2015). It is worth our while, then, to spend some time considering Fisher's use of the analogy before we turn to contemporary efforts to understand universality.

4.2.1 R. A. Fisher and Statistical Physics

As I have described more fully elsewhere (Pence, in press, ch. 4–6), Fisher's use of statistical physics arises in a profoundly complex historical context, in which he is responding to developments in mathematical evolution driven by the "biometrical school" led by Karl Pearson and W. F. R. Weldon, as well as the advances in genetics and theories of inheritance that flourished after the rediscovery of Mendel's work. It is unarguably the case, however, that Fisher takes up analogies between evolutionary theory and statistical mechanics with the explicit goal of producing a more general and abstract theory of natural selection than those that were current in the literature of the day.[37]

He deploys statistical physics in the service of theoretical generality in two ways, which, following Hodge (1992), we can already see clearly in one of

---

[37] As is by now well known, he also did so with the explicit goal of advancing eugenics. A discussion of those facets of his work lies well beyond my scope here, but I can strongly recommend to interested readers the work of Mazumdar (1992) and Moore (2007).

Fisher's earliest papers, a 1915 article on eugenics that he co-authored with C. S. Stock (Fisher and Stock, 1915). To build a first analogy between physics and evolution, Fisher considers the way in which macro-level quantities could arise from a population of micro-level individuals, and argues that evolution and statistical mechanics are alike in that a statistical property guarantees the stability of their results: "the reliability and predictability of the outcome" in both cases, as Hodge puts it, emerges "when the individuals are numerous and the causes acting upon them independent" (Hodge, 1992, p. 248). In a later 1922 paper on natural selection, Fisher would summarize this point by writing that "'the distribution of the frequency ratio' for different hereditary factors is – in the absence of selection and random survival effects and so on – a stable one like that of 'velocities in the Theory of Gases'" (Hodge, 1992, p. 249). It is the very fact that evolving systems form a statistical population that lets us draw certain kinds of generalized conclusions from them.

Fisher's second analogy with statistical mechanics invokes yet another element of universality. In physics, Fisher writes, we are "independent of particular knowledge about separate atoms, as in eugenics we are independent of particular knowledge about individuals" (Fisher and Stock, 1915, pp. 60–61). He remained worried about what might happen if this failed to be the case: In the introduction to his well-known 1918 paper, Fisher thought that the importance of "assumptions about the dominance of particular factors, about the size of their effects, about their proportion in the population, about dimorphism and polymorphism, and about linkage" (Hodge, 1992, pp. 248–249) might make it impossible to construct a general theory of natural selection. A few years later, he would argue that what was missing in evolutionary theory was an investigation like "the analytical treatment of the Theory of Gases, in which it is possible to make the most varied assumptions as to the accidental circumstances, and even the essential nature of the individual molecules, and yet to develop the general laws as to the behaviour of gases, leaving but a few fundamental constants to be determined by experiment" (Fisher, 1922, pp. 321–322).

That said, in his work from 1915 to 1930, Fisher appears to be confused on this point in at least two ways. First, it is never clear just what it is that we are abstracting away *from* in Fisher's theoretical framework. His Fundamental Theorem of Natural Selection, supposedly a central part of the analogy and playing the role of the Second Law of Thermodynamics, is phrased in two different ways in the *Genetical Theory*, once with respect to individuals and once with respect to species. In Turner's words, Fisher's discussions "repeatedly fail to make clear whether the 'organism' Fisher is discussing is a single individual or a whole population or even species" (Turner, 1987, p. 327).

We can, of course, reconstruct after the fact what Fisher must have had in mind in order to make his mathematics consistent, but this is not a ringing endorsement of the feasibility of Fisher's analogies.

Second, Fisher doesn't offer us an explanation of how and why evolutionary theory actually does have this status of universality, or what facts about evolutionary theory are relevant for guaranteeing that status. (As we will see below, this is in some sense unsurprising; the physical treatments of universality that would be most suitable for applying to biology date only from the second half of the twentieth century. Fisher's formal tools were thus likely not up to the task.) He also recognizes that even though the understanding of natural selection that he develops in the Fundamental Theorem happens to be independent from the lower-level details, this is a contingent feature of this model, and he is open to the possibility that future work would reveal that such independence could no longer be sustained. As he puts it,

> Generalized description should, however, never be regarded as an aim in itself. It is at best a means toward apprehending the causal processes which have given rise to the phenomena observed. Beyond a certain point it can only be pursued at the cost of omitting or ignoring real discrepancies of detail, which, if the causes were understood, might be details of great consequence. (Fisher, 1930, p. 178)

Fisher's approach to causation, in turn, would take us too far afield for my purposes here.[38] But straightforwardly reading Fisher's work leaves us with something of a contradiction – an unsteady balance between the attempt to derive a statistical theory of high generality and the grounding of that theory in the processes that give rise to population-level selection. In order to obtain further clarity on this question, it's time that we turn to the contemporary perspective on universality.[39]

4.2.2 Contemporary Statistical Physics and Universality

The classic approach in physics to understanding why statistical mechanics provides us with these kinds of generalizable results appeals to a property known as *ergodicity*.[40] Start as we did in Figure 5 by defining the state space of micro-level states and the state space of macro-level states of the system.

---

[38] See Moore (2007) for an excellent discussion of the relationship between Fisher's view of indeterministic causes, eugenics, and Christianity.

[39] In recent years there has also been increasing discussion and reinterpretation of Fisher's Fundamental Theorem of Natural Selection in the biological literature. An excellent summary of the issues in play can be found in Okasha (2008).

[40] My presentation here follows Sklar (1992, pp. 117–127).

We would like to claim that the important macro-level quantities (like entropy or temperature) are derived from the values of the important micro-level quantities. But in order to calculate their values (which will always involve, in some sense, "averaging" over all of the micro-states), we have to know how likely it is that we will find the micro-level constituents of the system in each of their possible states – a so-called "natural" probability measure over micro-level states.

The simplest way to do this would be to use some kind of principle of indifference, assigning equal probabilities to each of the possible states (or each set of possible states). But, recalling the arguments of van Fraassen (1989, ch. 12), we know that such putatively "objective" uses of a principle of indifference require further justification. In van Fraassen's classic example, as we inspect the output of cubes from a cube factory, we have to decide whether we think the probabilities are uniformly distributed that the factory will produce a cube with side-length l, or a cube with face-area a, or a cube with volume V – and each of these "indifferent" distributions will give us different predictions about the cubes that the factory produces. In the same way, we can count up micro-level states in a variety of different ways, grouping them together by different kinds of criteria, which will in turn produce different probability distributions after we assign them "equal" probabilities.

How do we justify the choice of the natural probability measure over micro-level states? One approach, pioneered by Boltzmann, takes advantage of the long-term behavior of systems in statistical mechanics. Call an individual system *ergodic* if it would, in the long run, pass through *every* micro-level state that it could reach from its initial conditions. For ergodic systems, Boltzmann proved that there is only *one* natural measure of probability for the micro-states: Namely, the probability of some micro-state is just the proportion of time that the system would spend in that state over the (infinite) long run. For ergodic systems, then, we have one single natural probability measure, and we can use that probability measure to ground the way in which we derive the macro-level quantities. We can use this measure and averaging process to abstract away from the micro-level details of every ergodic system.

There's only one problem: The hypothesis that any real-world system more complex than "a small number of perfectly elastic spheres in a box" is ergodic seems to be straightforwardly false. Physicists have since endeavored to find a more complex theoretical framework that might enable us to get a similarly robust, single natural probability measure without assuming that ergodicity holds in general. But these approaches are still incomplete and suffer from a number of conceptual worries. Perhaps most significantly, as Batterman notes, "since we do often witness systems that are not in equilibrium, it is difficult to maintain the identification of thermodynamical values with *infinite* time averages" (Batterman, 1998, p. 187). It's also notable that ergodicity would be even *more* straightforwardly false if we considered exporting it from physics to a potential analogy with evolutionary theory – no population could even conceivably visit *every* distribution of traits, no matter how long we leave it to evolve (developmental constraints, at the very least, would forbid that).

If the problems with this perspective are difficult to surmount, what replacement might we offer for the ergodic approach that could lead to a more general understanding of universality? I'll consider two possibilities here. One, as argued for by Sklar, rejects the very question and involves the abandonment of ergodic theory and universality entirely.

> But there is a much simpler answer. And it is correct. And it is the *full* answer.
> And it is totally independent of any ergodic results. It goes like this: How a
> gas behaves over time depends upon (1) its microscopic constitution; (2) the
> laws governing the interaction of its micro-constituents; (3) the constraints
> placed upon it; and (4) *the initial conditions characterizing the microstate of
> the gas at a given time.* (Sklar, 1973, p. 210, original emphasis)

In other words, we simply take the micro-level seriously and understand the way in which the initial conditions there can lead us to the kinds of macro-level outcomes that interest us. Rather than trying to find an overarching theoretical justification, we ought instead to recognize that it is the "matter-of-fact distribution of such initial conditions" (Sklar, 1973, p. 210) that is found among our real-world samples of gas that is responsible for our success in deriving macro-level explanations. The micro-level facts matter after all.

I will return to the evolutionary analogue of this explanation below; first, consider the other alternative, as laid out by Batterman. Drawing on the work of Khinchin (1949), he argues (in terms quite reminiscent of the statisticalist view) that both ergodic theory and Sklar's response have missed the point. Instead of giving an argument about particular ergodicity-replacing properties of thermodynamical systems, or the initial conditions of particular gas samples, what we must instead do is explain why "statistical mechanical arguments work *because of*, and not just in spite of, its 'complete abstraction' from the nature of the forces and types of interactions responsible for the actual motions of the systems being studied" (Batterman, 1998, p. 185). It is this "broader" explanation for the success of statistical mechanics for which we should be searching.

Let's see how Khinchin's proposal works, then consider how Batterman expands upon it. Recall that we were in the business of attempting to define the functions that describe the macro-level quantities on the basis of the micro-level quantities (in the ergodic case, these were weighted averages). Khinchin realized that such a result remains derivable in general if those functions have a very particular kind of structure – in short, if they represent macro-quantities as sums of some independent properties belonging to each of the micro-level elements taken in isolation (in his terms, "sum functions"). If this holds, then we can apply the central limit theorem and derive the expected normal distribution for those macro-level quantities (though see Lyon, 2014).

While this program could succeed for particular cases, it, too, suffers from major problems. First, it's not clear how many functions for macro-level quantities could be rewritten in the right way. Second, even for those that can, "the decomposability of the system into components in the sense just described excludes the possibility that the components interact with one another energetically" (Batterman, 1998, p. 193). Put differently, the sum-functions approach relies on a lack of relevant individual-level interactions. This assumption is implausible in statistical physics and would be even more so in evolutionary biology.[41]

Part of the problem here concerns behavior at what are known in statistical mechanics as *critical points* – the most common examples of these from everyday experience being phase transitions (like melting or boiling). In normal equilibrium behavior, distant parts of a system are not, or are only loosely, correlated with one another. When those correlations are absent, it becomes easier to imagine the parts of the system as noninteracting elements amenable to a Khinchin-style treatment. But "at the critical point, components of the system that are widely separated spatially become strongly correlated" (Batterman, 1998, p. 198). There are still universal phenomena that can be described at this level – a basic fact in statistical mechanics – but Khinchin's work (and other extensions of it) will not be applicable here.

Batterman's amendment to Khinchin, then, is to consider how to carry out his program with the theoretical apparatus that *does* describe behavior at critical points: the renormalization group. While describing this theory informally is relatively difficult, we can think of it as locating fixed points in the evolution of these systems – that is, the very universal behaviors that we are hoping to explain – and then describing what kinds of dynamics will continue to evolve to that same terminal fixed point. In short, it offers us a way to describe the stability of a system under the perturbation of certain kinds of facts about how it is put together:

---

[41] There are also a number of problems with his technical apparatus that lead to failures too complex for me to describe here; see Batterman (1998, pp. 196–198).

> This *stability* under perturbation demonstrates that certain facts about the microconstituents of the systems are individually largely irrelevant for the systems' behaviors at criticality. Instead, their *collective properties* dominate their critical behavior, and these collective properties are characterized by the fixed points of the renormalization group transformation.... (Batterman, 1998, p. 200)

Batterman goes on to argue that this kind of pattern could underlie explanations in the special sciences. An engineer would like to know where the "singularities" in her equation describing the load on a bridge are located (i.e., the configurations under which the bridge might collapse). But without an exact equation for this behavior (as a result of the difficulty of analyzing the materials), what she is really concerned with, Batterman writes, is "the equation's (structural) stability – roughly, the stability of the topology of its solutions under perturbation of its form" (Batterman, 1998, p. 200).[42]

At least in theory, this sort of approach is available in the evolutionary case. We know that the kinematic properties in population-level selection (that is, the trait fitnesses) are highly causally interdependent, and hence highly correlated with one another. (The utility of particular traits is often a function both of their background and of their direct interactions with other traits.) There is also, by analogy with other cases where renormalization-type explanations are available, a sort of "asymptotic approach" to the phenomena of interest, as the size of a population increases and the quirks of individuals become less likely to cause major changes in population-level quantities (Batterman, 2000, p. 129).

Foregrounding the potential irrelevance of individual-level facts to population-level phenomena of natural selection has, therefore, proven to be an interesting and provocative place for further philosophical study. Far from being a trivial move that results from a quick analysis of the characteristics of the relevant mathematical models, there is a rich and varied literature in the philosophy of physics endeavoring to understand just how these kinds of universal phenomena are supposed to come about in the case of statistical mechanics. My quick survey of that literature here leaves us with two approaches upon which we might draw. One, advocated by Sklar, appears much like an argument for individual-level causalism. The micro-level facts about biological individuals – the very ways in which they instantiate processes of selection and inheritance – would prove essential for understanding the emergence of long-term, stable processes of directional, selective change. (It's unclear in Sklar's case whether the resulting processes of change are *causal*,

---

[42] He also extends this analysis to multiple realizability in general in Batterman (2000).

and hence whether this analogy preserves the population-level causation that is still present in the individual-level causalist view; it is here that more work would remain to be done.)

The other would draw on Batterman's invocation of the renormalization group in support of statisticalism. This is also not a simple effort. Demonstrating that the formal quantities at issue are in fact stable under perturbations of various types involves some real mathematical work within the formalism of that theory (see, e.g., Batterman, 1998, pp. 203–205). That kind of work should indeed be possible in the evolutionary context, given the quality of the formal modeling apparatus available to us in contemporary evolutionary theory. As of yet, we don't have the analogue to this kind of formal work clearly described – but we can now see how we might approach it, following out the analogy with Batterman's work.

Once again, this evaluation is not decisive – there is no clear argument for or against the conceptual coherence of the irrelevance of individual-level facts for natural selection at the population level. On the contrary, it is, I think, extremely instructive that the Sklar–Batterman division here can be recognized in our competing approaches to evolution. That said, drawing an analogy with another old and well-studied problem in the philosophy of science can offer us a clearer way to understand the problem, theoretical resources that we can use to attack it, and the possibility for future collaboration.

5 Conclusion

The debate over the causal status of natural selection has entailed a significant, not to say bewildering, collection of different questions about the interpretation of evolutionary theory. These range from traditional issues in what we might call the conceptual foundations of evolution – What is natural selection? Are natural selection and genetic drift to be understood as processes or as outcomes? What is fitness? – to those as general as the role of abstraction in scientific explanation or the nature of supervenience and multilevel causal systems.

Positions in the conflict between causalism and statisticalism have tended to bundle together sets of answers to all these various questions. As evidenced by the persistence of the debate in general, there is something compelling about such an approach – such unified pictures of an understanding of evolutionary theory can be extremely attractive. But, on balance, I think the combined method's demerits outweigh its advantages. Separating these questions gives us the opportunity for significantly increased leverage against difficult philosophical problems, clarity in presentation, and more opportunities for engagement with issues outside the philosophy of evolutionary theory.

In this work, then, I hope to have demonstrated this by example. I've focused on the single question (complex though it may be) of the causal structure of natural selection, offering a new framework on which we can diagram putative relationships between micro- and macro-level quantities, the kinematic properties that describe their changes, and the dynamical principles that let us pick selection out from other kinds of change. Such a framework is readily amenable to the presentation of the statisticalist perspective, to a population-level causalist view (here a view inspired by the work of Abrams), and to an individual-level causalist view (derived from work by myself and Ramsey).

Reinterpreting these views in this way, in turn, allows us to pick out a variety of their philosophically interesting features. The statisticalist position shows itself to be quite reliant upon what we can now identify as the question of universality: the phenomenon that renders the description at the population level independent of that at the individual level (and, by extension, renders population-level selection noncausal as well). While I did not have the space to consider the question here in the detail that it deserves, Abrams's picture's reliance upon his interpretation of causal probability is marked out as crucial. As for the individual-causalist picture, we were able to clearly see how much work remains to be done; in particular, to focus on the account that remains to be offered of how the processes of selection that act on individual lineages and that act on populations should be related.

I then turned in the last section to two ways in which this new approach could make contact with broader problems in philosophy. I argued that the individual-level causalist might take advantage of some novel results in the metaphysics of natural kinds and composition due to Boyd and Wilson, which could help us understand the relationship between individual-level and population-level natural selection by invoking the relation of composition as part of our causal story. Finally, the question of the causal independence of the population level parallels a significant technical literature on the idea of universality in physics – by no means taken for granted as a straightforward fact about thermodynamical theory, but rather the target of an extensive formal apparatus.

Two more general morals are worth drawing. First, has the effort to separate the set of issues surrounding causation and individual-population relationships really succeeded, or have we just found different language in which to express the same positions with which we began in Section 1? While I can't claim immediate success on this score, I hope to have done some work here by way of demonstrating that much less of the logical space is in fact occupied with respect to these questions than one might have thought. An apparently straightforward, two-sided debate has been shown to depend upon answers to dozens of questions (and these only after we restricted our focus to concerns of causal structure). There are thus myriad different combinations of answers to such questions that could create a much richer philosophical state space (at the risk of a bad pun) than is represented by the state of play in the literature today. (If my own experience and my personal communications with interlocutors in this debate are any indication, a more overt recognition of this diversity in possible positions is welcome.) Whether or not those will result in viable interpretations of evolutionary theory is a matter for future work, but I am optimistic that perhaps this will offer us a way forward.

Finally, another part of the advantage of teasing apart the various questions at play in this debate is the realization that at least some of them are not distinctly biological. This is true on the one hand of the more strictly metaphysical questions that I have focused on here – concerns about causal structures and supervenience. It is equally true of the discussions of explanation that have sometimes been broached in the context of this debate, including the nature of abstraction and observer-dependence in explanations. While sometimes the literature in the philosophy of biology has given lip service to this point, drawing parallels to either more abstract examples or to relatively simplified comparisons with other sciences, there has been an unfortunate lack of engagement with philosophical work in those domains.

I intend this Element, then, to not only argue for a new way of understanding the causal structure of natural selection, but as an appeal by demonstration for more extensive collaboration between philosophers of biology and other parts of the discipline – philosophy of the other special sciences, general philosophy of science, and metaphysics, at the very least. To put the point negatively, positions in the philosophy of biology have implications for these other areas, whether we reflectively engage with their literature or no. But to put the point positively, we can gain interesting and potentially useful insight by virtue of such engagement.

I am acutely aware that this Element raises more questions than it resolves. It remains true that evolutionary theory demands extensive and careful philosophical interpretation, and adding these further resources doesn't serve to foreclose on any of the available interpretive options. But in each case, our view of the relevant questions has become clearer, the field of available possibilites brought into sharper focus, and the arguments for and against them can draw on further sources of comparison and intuition. That suffices, at least for my understanding of philosophical progress.

References

Abrams, M. (2009a). Fitness "kinematics": Biological function, altruism, and organism-environment development. *Biology & Philosophy* **24**(4): 487–504. DOI: https://doi.org/10.1007/s10539-009-9153-2.

Abrams, M. (2009b). What determines biological fitness? The problem of the reference environment. *Synthese* **166**(1): 21–40. DOI: https://doi.org/10.1007/s11229-007-9255-9.

Abrams, M. (2012a). Measured, modeled, and causal conceptions of fitness. *Frontiers in Genetics* **3**: 196. DOI: https://doi.org/10.3389/fgene.2012.00196.

Abrams, M. (2012b). Mechanistic probability. *Synthese* **187**(2): 343–375. DOI: https://doi.org/10.1007/s11229-010-9830-3.

Abrams, M. (2015). Probability and manipulation: Evolution and simulation in applied population genetics. *Erkenntnis* **80**(S3): 519–549. DOI: https://doi.org/10.1007/s10670-015-9784-4.

Andersen, H. (2018). Complements, not competitors: Causal and mathematical explanations. *British Journal for the Philosophy of Science* **69**(2): 485–508. DOI: https://doi.org/10.1093/bjps/axw023.

Ariew, A. and Ernst, Z. (2009). What fitness can't be. *Erkenntnis* **71**(3): 289–301. DOI: https://doi.org/10.1007/s10670-009-9183-9.

Ariew, A. and Lewontin, R. C. (2004). The confusions of fitness. *British Journal for the Philosophy of Science* **55**(2): 347–363. DOI: https://doi.org/10.1093/bjps/55.2.347.

Ariew, A., Rice, C., and Rohwer, Y. (2015). Autonomous-statistical explanations and natural selection. *British Journal for the Philosophy of Science* **66**(3): 635–658. DOI: https://doi.org/10.1093/bjps/axt054.

Ayala, F. J., Tracey, M. L., Barr, L. G., McDonald, J. F., and Pérez-Salas, S. (1974). Genetic variation in natural populations of five Drosophila species and the hypothesis of the selective neutrality of protein polymorphisms. *Genetics* **77**(2): 343.

Baker, A. (2005). Are there genuine mathematical explanations of physical phenomena? *Mind* **114**(454): 223–238.

Batterman, R. W. (1998). Why equilibrium statistical mechanics works: Universality and the renormalization group. *Philosophy of Science* **65**(2): 183–208. DOI: https://doi.org/10.1086/392634.

Batterman, R. W. (2000). Multiple realizability and universality. *British Journal for the Philosophy of Science* **51**(1): 115–145. DOI: https://doi.org/10.1093/bjps/51.1.115.

Beatty, J. H. (1987). Dobzhansky and drift: Facts, values and chance in evolutionary biology. In L. Krüger, G. Gigerenzer and M. S. Morgan (eds.), *The Probabilistic Revolution, Volume 2: Ideas in the Sciences*. Bradford Books, Cambridge, MA, pp. 271–311.

Beatty, J. H. (1992). Random drift. In E. F. Keller and E. A. Lloyd (eds.), *Keywords in Evolutionary Biology*. Harvard University Press, Cambridge, MA, pp. 273–281.

Beatty, J. H. (1993). The evolutionary contingency thesis. In G. Wolters and J. G. Lennox (eds.), *Concepts, Theories, and Rationality in the Biological Sciences: The Second Pittsburgh-Konstanz Colloquium in the Philosophy of Science*. University of Pittsburgh Press, Pittsburgh, pp. 45–81.

Beatty, J. H. and Finsen, S. (1989). Rethinking the propensity interpretation of fitness: A peek inside Pandora's box. In M. Ruse (ed.), *What the Philosophy of Biology Is: Essays for David Hull*. Kluwer Academic Publishers, Dordrecht, pp. 18–30.

Bhogal, H. (2020). Difference-making and deterministic chance. *Philosophical Studies* **178**(7): 2215–2235. DOI: https://doi.org/10.1007/s11098-020-015 38-4.

Bouchard, F. and Rosenberg, A. (2004). Fitness, probability and the principles of natural selection. *British Journal for the Philosophy of Science* **55**(4): 693–712. DOI: https://doi.org/10.1093/bjps/55.4.693.

Boyd, R. N. (2017). How philosophers "learn" from biology – reductionist and antireductionist "lessons." In D. L. Smith (ed.), *How Biology Shapes Philosophy: New Foundations for Naturalism*. Cambridge University Press, Cambridge, pp. 276–301.

Bradie, M. (1986). Assessing evolutionary epistemology. *Biology & Philosophy* **1**(4): 401–459. DOI: https://doi.org/10.1007/BF00140962.

Brandon, R. N. (1990). The concept of environment in the theory of natural selection. *In Adaptation and Environment*. Princeton University Press, Princeton, NJ, pp. 45–77.

Brandon, R. N. (2005). The difference between selection and drift: A reply to Millstein. *Biology & Philosophy* **20**(1): 153–170. DOI: https://doi.org/10 .1007/s10539-004-1070-9.

Brandon, R. N. (2006). The principle of drift: Biology's first law. *Journal of Philosophy* **103**(7): 319–335.

Chiu, L. (2019). Decoupling, commingling, and the evolutionary significance of experiential niche construction. In T. Uller and K. N. Laland (eds.), *Evolutionary Causation: Biological and Philosophical Reflections*. The Massachusetts Institute of Technology Press, Cambridge, MA, pp. 299–322.

Currie, A. (2019). *Scientific Knowledge and the Deep Past: History Matters.* Cambridge University Press, Cambridge.

Dieckmann, U. (1997). Can adaptive dynamics invade? *Trends in Ecology & Evolution* **5347**(4): 128–131.

Douglas, H. E. (2009). Reintroducing prediction to explanation. *Philosophy of Science* **76**(4): 444–463. DOI: https://doi.org/10.1086/648111.

Doulcier, G., Takacs, P., and Bourrat, P. (2020). Taming fitness: Organism-environment interdependencies preclude long-term fitness forecasting. *BioEssays* **43**(1): e2000157. DOI: https://doi.org/10.1002/bies .202000157.

Earnshaw, E. (2015). Evolutionary forces and the Hardy–Weinberg equilibrium. *Biology & Philosophy* **30**(3): 423–437. DOI: https://doi.org/ 10.1007/s10539-014-9464-9.

Elgin, M. and Sober, E. (2017). Popper's shifting appraisal of evolutionary theory. *HOPOS* **7**(1): 31–55. DOI: https://doi.org/10.1086/691119.

Endler, J. A. (1986). *Natural Selection in the Wild.* Princeton University Press, Princeton, NJ.

Eronen, M. I. (2015). Levels of organization: A deflationary account. *Biology & Philosophy* **30**(1): 39–58. DOI: https://doi.org/10.1007/s10539-014-9461-z.

Fisher, R. A. (1918). The correlation between relatives on the supposition of Mendelian inheritance. *Philosophical Transactions of the Royal Society of Edinburgh* **52**: 399–433.

Fisher, R. A. (1922). On the dominance ratio. *Proceedings of the Royal Society of Edinburgh* **42**: 321–341. DOI: https://doi.org/10 .1016/S0092-8240(05)80012-6.

Fisher, R. A. (1930). *The Genetical Theory of Natural Selection.* Clarendon Press, Oxford.

Fisher, R. A. and Stock, C. S. (1915). Cuénot on preadaptation: A criticism. *Eugenics Review* **7**(1): 46–61.

Galton, F. (1889). *Natural Inheritance.* Macmillan, London.

Gayon, J. (1998). *Darwinism's Struggle for Survival: Heredity and the Hypothesis of Natural Selection.* Cambridge University Press, Cambridge.

Gillespie, J. H. (1974). Natural selection for within-generation variance in offspring number. *Genetics* **76**(3): 601–606.

Godfrey-Smith, P. (2009). *Darwinian Populations and Natural Selection.* Oxford University Press, Oxford.

Guay, A. and Sartenaer, O. (2016). A new look at emergence. Or when *after* is different. *European Journal for Philosophy of Science* **6**(2): 297–322. DOI: https://doi.org/10.1007/s13194-016-0140-6.

Hacking, I. (1990). *The Taming of Chance.* Cambridge University Press, Cambridge.

Haufe, C. (2013). From necessary chances to biological laws. *British Journal for the Philosophy of Science* **64**(2): 279–295. DOI: https://doi.org/10.1093/bjps/axs001.

Haug, M. C. (2007). Of mice and metaphysics: Natural selection and realized population-level properties. *Philosophy of Science* **74**(4): 431–451.

Hitchcock, C. and Velasco, J. D. (2014). Evolutionary and Newtonian forces. *Ergo* **1**. DOI: https://doi.org/10.3998/ergo.12405314.0001.002.

Hodge, M. J. S. (1983). Darwin and the laws of the animate part of the terrestrial system (1835–1837): On the Lyellian origins of his zoonomical explanatory program. *Studies in History of Biology* **6**: 1–106.

Hodge, M. J. S. (1987). Natural selection as a causal, empirical, and probabilistic theory. In L. Krüger, G. Gigerenzer and M. S. Morgan (eds.), *The Probabilistic Revolution, Volume 2: Ideas in the Sciences.* Bradford Books, Cambridge, MA, pp. 233–270.

Hodge, M. J. S. (1992). Biology and philosophy (including ideology): A study of Fisher and Wright. In S. Sarkar (ed.), *The Founders of Evolutionary Genetics.* Kluwer Academic Publishers, Dordrecht, pp. 231–293.

Hull, D. L. (1986). On human nature. *PSA: Proceedings of the Biennial Meeting of the Philosophy of Science Association* **1986**(2): 3–13.

Huneman, P. (2010). Topological explanations and robustness in biological sciences. *Synthese* **177**(2): 213–245. DOI: https://doi.org/10.1007/s11229-010-9842-z.

Ismael, J. (2009). Probability in deterministic physics. *Journal of Philosophy* **106**(2): 89–108.

Ismael, J. (2011). A modest proposal about chance. *Journal of Philosophy* **108**(8): 416–442.

Jenkin, F. (1867). [Review of] The Origin of Species. *North British Review* **46**: 277–318.

Khinchin, A. I. (1949). *Mathematical Foundations of Statistical Mechanics.* Dover, New York.

Kim, J. (1993). The nonreductivist's troubles with mental causation. In J. Heil and A. Mele (eds.), *Mental Causation.* Oxford University Press, New York, pp. 189–210.

Kisdi, É. and Geritz, S. A. H. (2010). Adaptive dynamics: A framework to model evolution in the ecological theatre. *Journal of Mathematical Biology* **61**(1): 165–9. DOI: https://doi.org/10.1007/s00285-009-0300-9.

Lange, M. (1995). Are there natural laws covering particular biological species? *Journal of Philosophy* **92**(8): 430–451.

Lange, M. (2013). Really statistical explanations and genetic drift. *Philosophy of Science* **80**(2): 169–188. DOI: https://doi.org/10.1086/670323.

Lange, M. (2016). *Because without Cause: Non-Causal Explanations in Science and Mathematics*. Oxford University Press, Oxford.

Loux, M. J. (2006). *Metaphysics: A Contemporary Introduction*. 3rd ed. Routledge, New York.

Love, A. C. (2018). Individuation, individuality, and experimental practice in developmental biology. In O. Bueno, R.-L. Chen, and M. B. Fagan (eds.), *Individuation, Process, and Scientific Practice*. Oxford University Press, Oxford, pp. 165–191. DOI: https://doi.org/10.1093/oso/9780190 636814.003.0008.

Luque, V. J. (2016). Drift and evolutionary forces: Scrutinizing the Newtonian analogy. *Theoria* **31**(3): 397–410.

Lyon, A. (2011). Deterministic probability: Neither chance nor credence. *Synthese* **182**(3): 413–432. DOI: https://doi.org/10.1007/s11229-010-9750-2.

Lyon, A. (2014). Why are normal distributions normal? *British Journal for the Philosophy of Science* **65**(3): 621–649. DOI: https://doi.org/10.1093/bjps /axs046.

Machery, E. (2008). A plea for human nature. *Philosophical Psychology* **21**(3): 321–329. DOI: https://doi.org/10.1080/09515080802170119.

Matthen, M. (2009). Drift and "statistically abstractive explanation." *Philosophy of Science* **76**(4): 464–487. DOI: https://doi.org/ 10.1086/648063.

Matthen, M. and Ariew, A. (2002). Two ways of thinking about fitness and natural selection. *Journal of Philosophy* **99**(2): 55–83.

Matthen, M. and Ariew, A. (2009). Selection and causation. *Philosophy of Science* **76**(2): 201–224. DOI: https://doi.org/10.1086/648102.

Mayr, E. (1961). Cause and effect in biology. *Science* **134**(3489): 1501–1506.

Mayr, E. (1976). Typological versus population thinking. *Evolution and the Diversity of Life: Selected Essays*. Belknap, Cambridge, MA, pp. 26–29.

Mazumdar, P. M. H. (1992). Ideology and method: R. A. Fisher and research in eugenics. *Eugenics, Human Genetics, and Human Failings: The Eugenics Society, Its Sources and Its Critics in Britain*. Routledge, London, pp. 96–145.

McLoone, B. (2018). Why a convincing argument for causalism cannot entirely eschew population-level properties: Discussion of Otsuka. *Biology & Philosophy* **33**(1–2): 11. DOI: https://doi.org/10.1007/s10539-018-9620-8.

McShea, D. W. and Brandon, R. N. (2010). *Biology's First Law: The Tendency for Diversity and Complexity to Increase in Evolutionary Systems*. University of Chicago Press, Chicago.

Merlin, F. (2016). Weak randomness at the origin of biological variation: The case of genetic mutations. In G. Ramsey and C. H. Pence (eds.), *Chance in Evolution*. University of Chicago Press, Chicago, pp. 176–195.

Mills, S. K. and Beatty, J. H. (1979). The propensity interpretation of fitness. *Philosophy of Science* **46**(2): 263–286. DOI: https://doi.org/10.1086/28 8865.

Millstein, R. L. (2002). Are random drift and natural selection conceptually distinct? *Biology & Philosophy* **17**(1): 33–53. DOI: https://doi.org/10.1023/A:1012990800358.

Millstein, R. L. (2005). Selection vs. drift: A response to Brandon's reply. *Biology & Philosophy* **20**(1): 171–175. DOI: https://doi.org/10.1007/s10539-004-6047-1.

Millstein, R. L. (2006). Natural selection as a population-level causal process. *British Journal for the Philosophy of Science* **57**(4): 627–653. DOI: https://doi.org/10.1093/bjps/axl025.

Millstein, R. L. (2008). Distinguishing drift and selection empirically: "The Great Snail Debate" of the 1950s. *Journal of the History of Biology* **41**(2): 339–367. DOI: https://doi.org/10.1007/s10739-007-9145-5.

Millstein, R. L. (2013). Natural selection and causal productivity. In H.-K. Chao, S.-T. Chen and R. L. Millstein (eds.), *Mechanism and Causality in Biology and Economics*. Springer, New York, pp. 147–163.

Millstein, R. L. (2016). Probability in biology: The case of fitness. In A. Hájek and C. Hitchcock (eds.), *The Oxford Handbook of Probability and Philosophy*. Oxford University Press, Oxford, pp. 601–622.

Millstein, R. L., Skipper, R. A., and Dietrich, M. R. (2009). (Mis)interpreting mathematical models: Drift as a physical process. *Philosophy and Theory in Biology* **1**: e002.

Moore, J. (2007). R. A. Fisher: A faith fit for eugenics. *Studies in History and Philosophy of Biological and Biomedical Sciences* **38**(1): 110–135. DOI: https://doi.org/10.1016/j.shpsc.2006.12.007.

Nicholson, D. J. (2012). The concept of mechanism in biology. *Studies in History and Philosophy of Biological and Biomedical Sciences* **43**(1): 152–163. DOI: https://doi.org/10.1016/j.shpsc.2011.05.014.

Nicholson, D. J. (2014). The return of the organism as a fundamental explanatory concept in biology. *Philosophy Compass* **9**(5): 347–359. DOI: https://doi.org/10.1111/phc3.12128.

Okasha, S. (2006). *Evolution and the Levels of Selection*. Clarendon Press, Oxford.

Okasha, S. (2008). Fisher's fundamental theorem of natural selection—a philosophical analysis. *British Journal for the Philosophy of Science* **59**(3): 319–351. DOI: https://doi.org/10.1093/bjps/axn010.

Otsuka, J. (2016). Causal foundations of evolutionary genetics. *British Journal for the Philosophy of Science* **67**(1): 247–269. DOI: https://doi.org/10.1093/bjps/axu039.

Otsuka, J. (2019). *The Role of Mathematics in Evolutionary Theory*. Cambridge University Press, Cambridge.

Otsuka, J., Turner, T., Allen, C., and Lloyd, E. A. (2011). Why the causal view of fitness survives. *Philosophy of Science* **78**(2): 209–224. DOI: https://doi.org/10.1086/659219.

Pence, C. H. (2017). Is genetic drift a force? *Synthese* **194**(6): 1967–1988. DOI: https://doi.org/10.1007/s11229-016-1031-2.

Pence, C. H. (2018). Sir John F. W. Herschel and Charles Darwin: Nineteenth-century science and its methodology. *HOPOS* **8**(1): 108–140. DOI: https://doi.org/10.1086/695719.

Pence, C. H. (in press). *The Rise of Chance in Evolutionary Theory: A Pompous Parade of Arithmetic*. Elsevier.

Pence, C. H. (2021). W. F. R. Weldon changes his mind. *European Journal for Philosophy of Science* **11**(3): 1–20. DOI: https://doi.org/10.1007/s13194-021-00384-3.

Pence, C. H. and Ramsey, G. (2013). A new foundation for the propensity interpretation of fitness. *British Journal for the Philosophy of Science* **64**(4): 851–881. DOI: https://doi.org/10.1093/bjps/axs037.

Pence, C. H. and Ramsey, G. (2015). Is organismic fitness at the basis of evolutionary theory? *Philosophy of Science* **82**(5): 1081–1091. DOI: https://doi.org/10.1086/683442.

Pfeifer, J. (2005). Why selection and drift might be distinct. *Philosophy of Science* **72**(5): 1135–1145. DOI: https://doi.org/10.1086/508122.

Pigliucci, M. (2007). Do we need an extended evolutionary synthesis? *Evolution* **61**(12): 2743–2749. DOI: https://doi.org/10.1111/j.1558-5646.2007.00246.x.

Plutynski, A. (2007). Drift: A historical and conceptual overview. *Biological Theory* **2**(2): 156–167. DOI: https://doi.org/10.1162/biot.2007.2.2.156.

Popper, K. R. (1974). Darwinism as a metaphysical research programme. In P. A. Schilpp (ed.), *The Philosophy of Karl Popper*. Open Court Press, New York and Chicago, pp. 133–143.

Ramsey, G. (2006). Block fitness. *Studies in History and Philosophy of Biological and Biomedical Sciences* **37**(3): 484–498. DOI: https://doi.org/10.1016/j.shpsc.2006.06.009.

Ramsey, G. (2013a). Can fitness differences be a cause of evolution? *Philosophy and Theory in Biology* **5**(1): e401.

Ramsey, G. (2013b). Driftability. *Synthese* **190**(17): 3909–3928. DOI: https://doi.org/10.1007/s11229-012-0232-6.

Ramsey, G. (2013c). Organisms, traits, and population subdivisions: Two arguments against the causal conception of fitness? *British Journal for the Philosophy of Science* **64**(3): 589–608. DOI: https://doi.org/10.1093/bjps/axs010.

Ramsey, G. (2016). The causal structure of evolutionary theory. *Australasian Journal of Philosophy* **94**(3): 421–434. DOI: https://doi.org/10.1080/00048402.2015.1111398.

Reisman, K. and Forber, P. (2005). Manipulation and the causes of evolution. *Philosophy of Science* **72**(5): 1113–1123. DOI: https://doi.org/10.1086/508120.

Rice, S. H. (2004). *Evolutionary Theory: Mathematical and Conceptual Foundations*. Sinauer Associates, Sunderland, MA.

Sedgwick, A. (1860). Objections to Mr. Darwin's theory of the origin of species. *The Spectator* **33**: 285–286.

Shapiro, L. and Sober, E. (2007). Epiphenomenalism – the dos and the don'ts. In G. Wolters and P. Machamer (eds.), *Thinking about Causes: From Greek Philosophy to Modern Physics*. University of Pittsburgh Press, Pittsburgh, pp. 235–264.

Sheynin, O. B. (1980). On the history of the statistical method in biology. *Archive for History of Exact Sciences* **22**(4): 323–371.

Sklar, L. (1973). Statistical explanation and ergodic theory. *Philosophy of Science* **40**(2): 194–212.

Sklar, L. (1992). *Philosophy of Physics*. Westview Press, Boulder, CO.

Skow, B. (2011). Does temperature have a metric structure? *Philosophy of Science* **78**(3): 472–489. DOI: https://doi.org/10.1086/660304.

Sloan, P. R. (2009). The making of a philosophical naturalist. In M. J. S. Hodge and G. Radick (eds.), *The Cambridge Companion to Darwin*, 2nd ed. Cambridge University Press, Cambridge, pp. 21–43.

Sober, E. (1984). *The Nature of Selection*. The Massachusetts Institute of Technology Press, Cambridge, MA.

Sober, E. (2001). The two faces of fitness. In R. S. Singh (ed.), *Thinking about Evolution: Historical, Philosophical, and Political Perspectives*. The Massachusetts Institute of Technology Press, Cambridge, MA, pp. 309–321.

Sober, E. (2008). *Evidence and Evolution: The Logic behind the Science*. Cambridge University Press, Cambridge.

Sober, E. (2013). Trait fitness is not a propensity, but fitness variation is. *Studies in History and Philosophy of Biological and Biomedical Sciences* **44**(3): 336–341. DOI: https://doi.org/10.1016/j.shpsc.2013.03.002.

Stephens, C. (2004). Selection, drift, and the "forces" of evolution. *Philosophy of Science* **71**(4): 550–570. DOI: https://doi.org/10.1086/423751.

Sterelny, K. (1996). The return of the group. *Philosophy of Science* **63**(4): 562–584. DOI: https://doi.org/10.1086/289977.

Sterrett, S. G. (2002). Darwin's analogy between artificial and natural selection: How does it go? *Studies in History and Philosophy of Biological and Biomedical Sciences* **33**(1): 151–168. DOI: https://doi.org/10.1016/S1369-8486(01)00039-5.

Stoltzfus, A. (2019). Understanding bias in the introduction of variation as an evolutionary cause. In T. Uller and K. N. Laland (eds.), *Evolutionary Causation: Biological and Philosophical Reflections*. The Massachusetts Institute of Technology Press, Cambridge, MA, pp. 29–61.

Stoltzfus, A. and Yampolsky, L. Y. (2009). Climbing mount probable: Mutation as a cause of nonrandomness in evolution. *Journal of Heredity* **100**(5): 637–647. DOI: https://doi.org/10.1093/jhered/esp048.

Street, S. (2006). A Darwinian dilemma for realist theories of value. *Philosophical Studies* **127**(1): 109–166.

Strevens, M. (2011). Probability out of determinism. In C. Beisbart and S. Hartmann (eds.), *Probabilities in Physics*. Oxford University Press, Oxford, pp. 339–364.

Strevens, M. (2013). *Tychomancy: Inferring Probability from Causal Structure*. Harvard University Press, Cambridge, MA.

Strevens, M. (2016). The reference class problem in evolutionary biology: Distinguishing selection from drift. In G. Ramsey and C. H. Pence (eds.), *Chance in Evolution*. University of Chicago Press, Chicago, pp. 145–175.

Sturgeon, N. L. (1992). Nonmoral explanations. *Philosophical Perspectives* **6**(Ethics): 97–117.

Triviño, V. and Nuño de la Rosa, L. (2016). A causal dispositional account of fitness. *History and Philosophy of the Life Sciences* **38**: 6. DOI: https://doi.org/10.1007/s40656-016-0102-5.

Turner, J. R. G. (1987). Random genetic drift, R. A. Fisher, and the Oxford School of ecological genetics. In L. Krüger, G. Gigerenzer and M. S. Morgan (eds.), *The Probabilistic Revolution, Volume 2: Ideas in the Sciences*. Bradford Books, Cambridge, MA, pp. 313–354.

Uller, T. and Laland, K. N. (eds.) (2019). *Evolutionary Causation: Biological and Philosophical Reflections*. The Massachusetts Institute of Technology Press, Cambridge, MA.

van Fraassen, B. C. (1989). *Laws and Symmetry*. Clarendon Press, Oxford.

Walsh, D. M. (2003). Fit and diversity: Explaining adaptive evolution. *Philosophy of Science* **70**(2): 280–301. DOI: https://doi.org/10.1086/375468.

Walsh, D. M. (2007). The pomp of superfluous causes: The interpretation of evolutionary theory. *Philosophy of Science* **74**(3): 281–303. DOI: https://doi.org/10.1086/520777.

Walsh, D. M. (2010). Not a sure thing: Fitness, probability, and causation. *Philosophy of Science* **77**(2): 147–171. DOI: https://doi.org/10.1086/651320.

Walsh, D. M. (2013). Descriptions and models: Some responses to Abrams. *Studies in History and Philosophy of Biological and Biomedical Sciences* **44**(3): 302–308. DOI: https://doi.org/10.1016/j.shpsc.2013.06.004.

Walsh, D. M., Ariew, A. and Matthen, M. (2017). Four pillars of statisticalism. *Philosophy, Theory, and Practice in Biology* **9**: 1. DOI: https://doi.org/10.3998/ptb.6959004.0009.001.

Walsh, D. M., Lewens, T. and Ariew, A. (2002). The trials of life: Natural selection and random drift. *Philosophy of Science* **69**(3): 429–446. DOI: https://doi.org/10.1086/342454.

Weldon, W. F. R. (1893). On certain correlated variations in *Carcinus mœnas*. *Proceedings of the Royal Society of London* **54**: 318–329. DOI: https://doi.org/10.1098/rspl.1893.0078.

Wilson, J. M. (2010). From constitutional necessities to causal necessities. In H. Beebee (ed.), *The Semantics and Metaphysics of Natural Kinds*. Routledge, New York, pp. 192–211.

Woodward, J. (2003). *Making Things Happen*. Oxford University Press, Oxford.

Acknowledgments

Various pieces of the work presented here have been percolating, written up, discarded, critiqued, rejected, and presented for almost as long as I have been working in the philosophy of biology, so it is very hard for me to appropriately pay my intellectual debts here. I should begin by offering my sincere appreciation to Marshall Abrams, with whom I've chatted about many of these issues over the years and who prepared an extremely extensive and gracious review of this work. My thanks as well to another extremely helpful anonymous review, and to comments from Brian McLoone and Victor Luque. I also owe thanks to Grant Ramsey, who first got me involved with debates on fitness, and to André Ariew, who has remained a tireless and charitable interlocutor despite our many differences. Finally, I received excellent commentary from an audience at ISHPSSB 2019 in Oslo, particularly Marshall Abrams, Gregory Radick, Yafeng Shan, Mike Buttolph, and Stephen Hecht Orzack; an audience at the Universidad Nacional de Educación a Distancia in Madrid, especially Davide Vecchi, Victor Luque, Diego Rasskin-Gutman, and David Teira; and a reading group at KU Leuven, especially Grant Ramsey, Bendik Aaby, James DiFrisco, Alejandro Gordillo-García, and Cristina Villegas.

Cambridge Elements ⸗

Philosophy of Biology

Grant Ramsey

KU Leuven

Grant Ramsey is a BOFZAP research professor at the Institute of Philosophy, KU Leuven, Belgium. His work centers on philosophical problems at the foundation of evolutionary biology. He has been awarded the Popper Prize twice for his work in this area. He also publishes in the philosophy of animal behavior, human nature and the moral emotions. He runs the Ramsey Lab (theramseylab.org), a highly collaborative research group focused on issues in the philosophy of the life sciences.

Michael Ruse

Florida State University

Michael Ruse is the Lucyle T. Werkmeister Professor of Philosophy and the Director of the Program in the History and Philosophy of Science at Florida State University. He is Professor Emeritus at the University of Guelph, in Ontario, Canada. He is a former Guggenheim fellow and Gifford lecturer. He is the author or editor of over sixty books, most recently *Darwinism as Religion: What Literature Tells Us about Evolution; On Purpose; The Problem of War: Darwinism, Christianity, and their Battle to Understand Human Conflict;* and *A Meaning to Life.*

About the Series

This Cambridge Elements series provides concise and structured introductions to all of the central topics in the philosophy of biology. Contributors to the series are cutting-edge researchers who offer balanced, comprehensive coverage of multiple perspectives, while also developing new ideas and arguments from a unique viewpoint.

Cambridge Elements ≡

Philosophy of Biology

Elements in the Series

The Missing Two-Thirds of Evolutionary Theory
Robert N. Brandon and Daniel W. McShea

Reduction and Mechanism
Alex Rosenberg

Inheritance Systems and the Extended Evolutionary Synthesis
Eva Jablonka and Marion J. Lamb

How to Study Animal Minds
Kristin Andrews

Model Organisms
Rachel A. Ankeny and Sabina Leonelli

Comparative Thinking in Biology
Adrian Currie

Social Darwinism
Jeffrey O'Connell and Michael Ruse

Adaptation
Elisabeth A. Lloyd

Stem Cells
Melinda Bonnie Fagan

The Metaphysics of Biology
John Dupré

Facts, Conventions, and the Levels of Selection
Pierrick Bourrat

The Causal Structure of Natural Selection
Charles H. Pence

A full series listing is available at www.cambridge.org/EPBY

Printed in the United States
by Baker & Taylor Publisher Services